AF654624

Agricultural Information Systems and Services: Changes and Challenges

Agricultural Information Systems and Services: Changes and Challenges

Dr. K. Veeranjaneyulu

University Librarian

ANGR Agricultural University

Hyderabad

BSP **BS Publications**

A unit of **BSP Books Pvt. Ltd.**

4-4-309/316, Giriraj Lane, Sultan Bazar,
Hyderabad - 500 095 - A.P.

© 2012, *by Publisher*

First Reprint, 2018

All rights reserved

No part of this book or parts thereof may be reproduced, stored in a retrieval system or transmitted in any language or by any means, electronic, mechanical, photocopying, recording or otherwise without the prior written permission of the author.

Published by :

BS Publications
A unit of **BSP Books Pvt. Ltd.**

4-4-309/316, Giriraj Lane, Sultan Bazar,
Hyderabad - 500 095 - A.P.
Phone : 040 - 23445688, 23445600
e-mail : info@bspbooks.net
www.bspublications.net

Printed and bound in India by Bhavish Graphics.

ISBN : 978-93-81075-76-0 (Paperback)

Dedicated to

Dr. C. Devakumar

An Eminent Academician,
Scientist and Administrator
With love and respects

Foreword

"Nature is God's garb"
- Hopkins

This book on "Agricultural Information Systems and Services: Changes and Challenges" is a compendium of articles written over a period of time by a dashing Information Scientist Dr. K. Veeranjaneyulu, University Librarian, Acharya N.G. Ranga University, Hyderabad. I feel proud to write the foreword for the book brought out in honour of Dr. C. Devakumar, a distinguished Scientist-cum-Administrator.

Agricultural was the earliest occupation of the human kind. It was the first culture of all cultures. Agriculture prospered in all ancient cultures of Egypt, India, Babylonia, China etc. It has prime place in the survival and prosperity of people in the past and present Civilizations. Agriculture is an important subject since it envelops in its orbit subjects such as Land, Crops both foodgrains and commercial, fertilizers, plant protection, horticulture, animal husbandry, poultry, dairy, fisheries, agriculture education, agricultural marketing and agricultural engineering. There are many other allied sub-subjects besides the above major ones which need information.

India inherited a richest agricultural heritage. Nearly 80% of the people of India are involved in agriculture. It contributes 26 per cent of Gross Domestic Product (GDP) and accounts for 18% share of the total value of the India's exports. It provides major input in our national economy. Our agriculture is now on the verge of Second Green Revolution due to the thrust provided in our Five Year plans.

Indian Agriculture sector needs maximum Information inputs for further development and prosperity. The information generated and disseminated should reach the farmers and the rural folk in their own language for their understanding and implementation. Most of our agricultural information is still English based and it is not understood by the users. Mass media should be intensively used in regional languages for the benefit of the people involved in agriculture. Local librarians should use digital technology and the Internet Connectivity to provide information to those farmers who need it in appropriate form. All the Agricultural libraries in the Country irrespective of category should develop a consortia for free flow of information from Kashmir to

Kanya Kumari. Every paisa spent on the generation and dissemination of information and ideas in agriculture will enrich the country thousand fold.

The articles in the collection have merit and good suggestions for working librarians in Agricultural University, College and Research Centres and to students, scholars & teachers as well in the field. The book deserves to be kept on the shelves of their libraries. Finally, I congratulate Dr. Veeranjaneyulu for his efforts in bring out this volume.

Sankranti

15-01-2012

Prof. L.S. Ramaiah
University Librarian (Rtd.)
CIEFL (now known as EFLU)
Hyderabad (A.P.)

Acknowledgements

There are many friends who are stalwarts and seniors to me in LIS profession all over India. It is a very long list and it is difficult for me to enumerate all of them for want of space. Among them I wish to extend my sincere thanks to Late Prof. N. Guruswamy Naidu, Prof. L.S. Ramaiah, Prof. T. Rajagopal, Prof. D. Chandran, Prof. B. Ramesh Babu, Dr. M. Koteswara Rao, Prof. Prem Singh, Dr. D. B. Eswara Reddy, Prof. P.S.G. Kumar, Dr. Sonal Singh and Dr. Stanley Madan Kumar for their professional encouragement all through my career.

I am grateful to my colleagues Dr. LSRCV Ramesh, Ms. Y. Uma Devi, Mr. T. Sreenivasa Rao and Mr. N. P. Ravi Kumar for their help and encouragement in publishing this volume. They were also co-authors for few of the articles in this collection.

It is not out of place if I particularly extend my heart-felt regards to Prof. L.S. Ramaiah for his suggestions and help in shaping the present publication. It is he who suggested that a publication on Agricultural Information Systems and Services is urgently needed in view of the multiple changes that are taking place in Agriculture Development in India. I also thank him for writing the Foreword to this volume.

I express my thanks to Mr. Nikhil Shah and Mr. Anil Shah of M/S. B S Publications, Hyderabad for their interest, help and encouragement in publishing the Library and Information Science books. They are now the leading publishers in this region. They brought out this title with excellent design with in a short time maintaining all standards of a good publication.

Sankranthi
Hyderabad
15-01-2012

Dr. K. VEERANJANEYULU

Contents

AN ASSESSMENT OF AGRICULTURAL INFORMATION TECHNOLOGY IN INDIA : THE PRESENT STATUS

INTRODUCTION

Today India's population is estimated to be nearly a 1,000 million. This growing population is putting tremendous pressure on the food grain production in the country. India would need to produce an additional 100 million tons of food grains by 2020 AD to feed its anticipated population of more than one billion. Despite the great increase in the productivity of farm as a result of green revolution, there is marked imbalance between population pressure and adequate availability and utilization of resources of food. To make agriculture not only an instrument of food security but also livelihood security of majority of population, requires investment in the development and dissemination of techniques (Swaminathan, 1997).

The present rate of agricultural production could be doubled if the available technologies are brought to bear with the production process and programmes. Focusing more and more on transferring our new technologies away from the confines of laboratories and research institutes to the farmers and making them more result and work oriented is the need of the day. Since 90 per cent of the farmers are either functionally illiterate, information and technologies generated in laboratories, research stations cannot be availed by them. The information can be disseminated through extension workers.

AGRICULTURAL SITUATION IN INDIA

It is very often heard that the agricultural research in India is 'Ivory tower' research. This is because of the fact that the work of the research had very little impact on either the practice adopted by the farmers or on productivity. In retrospect it is clear that if research had little impact on the actual farming conditions, the fault cannot be attributed solely to the Scientists. This dilemma is directly related to the principle of "In out" supply of the needs of farmers and the "product demand" of the produce, taking the advantage of scientific findings. There is a plenty of "know how" of agricultural technologies of the developed countries and many internal agencies like FAO, CGIAR, ACIAR, ICAR and a host of other organizations have enlarged their network to extend, promote, identify the resources to accomplish the food problems in India.

CHANGING ENVIRONMENT

Agricultural technology has developed more rapidly in the 20th century than in all the previous times. The application of genetics to developing new strains of plants and animals has brought major changes in agriculture since 1920's.

It is obvious that research and development in both the private and public sectors have been playing an increasingly important role in shaping Agricultural technology. Utilization of renewable resources in agricultural crops is another area where biotechnology is playing a vital role and revolutionary advances in genetic engineering have come very handy for such utilization.

TECHNOLOGY GENERATION, ITS TRANSFER AND LINKAGE

Within the network of national agricultural research system, the two-wing research and extension can be visualized as operating linkages between polices and production levels. For effective functioning the following components need special consideration.

- Technology generating
- Technology transfer,
- Technology using sub systems.

TECHNOLOGY ASSESSMENT THROUGH NETWORKS

In addition to Internet, e-mail, fax services and technology based systems in operation play a vital role in agricultural technology assessment.

Moreover, it is not difficult to envisage all the components of IT used in agricultural research and extension institutions in a developing country like India.

The progress of research in agricultural fields and improvement essentially require timely supply of information to agricultural scientists. It is not possible to achieve self-sufficiency in food production without adequate information supply to agricultural scientists.

The concentration of effort is probably efficacious in informing agricultural scientists of improved methods in agriculture. It also indicates an unfortunate parochialism affecting both the agricultural scientists and the information handling industry. To promote an effective working relationship between research workers, who generate new agricultural knowledge, and the information and industry, information systems act as "middlemen" in a position to deliver such knowledge to the required points of distribution. The sources of agricultural information are varied in kind and widely scattered in location with an equally widespread distribution. In the same way agricultural information users are also equally scattered. To bridge these two, we need agricultural information networks.

Keeping this in mind a project called Agricultural Research Information system (ARIS) is being implemented with the financial aid of World Bank.

Collaboration among scientists in India, systematic access to research information available in India as well as in other countries, better project management of agricultural research and modernization of the office automation tools used by the scientists and the administrators are major goals of the ARIS project.

It is envisaged that point to point connectivity will be provided to all SAUs, ICAR Institutes, project directorates, national research centers and zonal research stations, through a WAN using dialup, leased lines and VSATs.

Besides creating the infrastructural facilities, ARIS have five information modules namely.

1. Agricultural Management Information System (ARMIS)
2. Agricultural Personal Information System (ARPIS)
3. Agricultural Financial Information System (ARFIS)
4. Agricultural Library Information System (ARLIS)
5. Training of Man power in field of IT.

ARIS has successfully accomplished the task of generating a wide interest for IT and use of computers among agricultural scientists, task of point to point connectivity for E-mail to 63 ICAR institutes, conducted several computer training programmes in the areas of data communication and networking, PC trouble shooting, MS window environment and procurement of computer and network equipment for enhancing the connectivity through WAN.

CONCLUSION

IT now impinges on virtually every aspect of society, and has developed dramatically at both extremes of its use. For all areas of science, IT is now of immense importance; both as research tool and as a medium of facilitating the information transfer process. These trends and developments which are particularly relevant to the transfer of scientific information through projects like ARIS is a boon to Agricultural Research Community.

While discussing the present status of agricultural information technology in India it is the need of the hour to adopt the newer technologies from time to time so as to minimize the drawbacks of the past to march to step forward towards new millennium. We hope India would certainly maintain the status of largest producer of food grains, pulses etc. by updating the information technology packages. As the population is crossing 100 crores, it is suggested to employ new technologies as and when they come into play.

REFERENCES

1. Contant, R, B, and Bottomley, A. (1986). Priority setting in Agricultural Research. International Service for National Agricultural Research working paper no. 10. Haque, Netherlands. Kalpage FS.C.P. 1977 Technical information services in developing countries. Requirement in information service for agriculture. *Proceedings of the International Conference COSTED,* Bangalore, March, 1977.

2. Phadinis, S.P.; Shoab, A. and Vijaya Lakshmi. N.R. (1971). Training of Agricultural Scientists in Information use. *Journal of Library Science* 2(2), pp. 151-181.

3. Ramesh et. *al.* (1999). The ARIS Project - Modern Way to explore Agricultural Information System (Accepted for Publication).

4. Trigo, E. (1986). Agricultural Research Organization in the developing world diversity and evaluation. *ISNAR Working Paper* No.4. Haque, Netherlands.

NEED FOR CONSORTIA FOR AGRICULTURAL INFORMATION IN INDIA

INTRODUCTION

The progress of research in agricultural fields and improvement of production essentially requires timely supply of information to agricultural scientists. It is not possible to achieve self-sufficiency in food production without adequate information supply to agricultural scientists. The sources of agricultural information are varied in kind and widely scattered in location with an equally widespread distribution. In the same way agricultural information users are also equally scattered. To bridge this gap, we need Agricultural Information System. The above have made it imperative to share library resources and to go for consortia of agricultural libraries in India.

WHAT IS LIBRARY CONSORTIA?

A consortium is a group of libraries that want to have a central place to store content in centralized manner and distribute to all associated libraries. Consortia is a generic term to indicate a group of libraries that are working together for a common goal to expand co-operation on traditional library services such as collection and development of electronic media. The aim of the consortia is to achieve what the members of the group cannot achieve individually.

Groups of Libraries come together with common interest to fulfill their requirement by joining hands. One library or agency can act as coordinator for identification of resources, managing consortium services, managing affairs related to common interest, implementation of its objective etc.

CONSORTIA INITIATIVE IN INDIA

The development of consortia in India is now at the stage of infancy. Several library consortia have been setup over last -few years to obtain licenses and to enable desktop internet access to scholarly electronic resources like e-journals and databases. Some of the major consortia formed in India over the years are:

FORSA libraries Consortia

Forum for Resource Sharing in Astronomy & Astrophysics (FORSA) was established in 1981. A Group was formed for the sharing of resources in areas common to research and development for group members. FORSA is subscribing to common database full text journals on behalf of all member libraries for their users.

Members of FORSA are Indian Institute of Astrophysics (IIA), Inter-University Centre for Astronomy & Astrophysics (IUCAA), National. Centre for Radio Astrophysics (NCRA), Physical Research Laboratory (PRL), Raman Research Institute (RRI), Tata Institute of Fundamental Research (TIFR), Nizamiah Observatory (NO) and Uttar Pradesh State Observatory (UPSO).

CSIR Consortia

Council of Scientific and Industrial Research (CSIR), was established in 1942. It has 40 research laboratories spread allover the country. More than 22,000 persons are working as scientists, researchers, and scholars. NISCAIR (earlier known as INSDOC -Indian National Scientific Documentation Centre, New Delhi and NISCOM -National Institute of Science Communication, New Delhi) on behalf of CSIR has initiated consortium "Consortium for CSIR Laboratories for Accessing e-journals". This consortium is negotiating with vendors for better pricing.

INDEST

The Ministry of Human Resource Development (MHRD), Govt. of India set up a consortium named as "Indian National Digital Library in Science and Technology (INDES'T) Consortium". This consortium was formed based on recommendations made by the group of experts appointed by the Ministry for Consortia-based Subscription to Electronic Resources for Technical Education System in India. Institutes like IISc, IITs, NITs, IIMs which are directly financially supported by MHRD are the regular members, Its membership is also open to other institutions.

UGC-Infonet

UGC-Internet E-Journals Consortium has been set up by the UGC to promote the use of electronic databases and full text access to journals by the Research and Academic community in the country. The Faculty, Research Scholars and Students of universities covered under UGC are the primary beneficiaries. This scheme will also be extended to colleges very soon.

UGC will bear the entire expenses for UGC funded universities for providing e-journals access on behalf of participating universities INFLIBNET Centre will subscribe resources based on the recommendations of National Negotiating Committee set up by UGC. Consortium covers all the disciplines viz. Pure Sciences, Social Sciences and Humanities including Management and Languages.

BENEFITS OF CONSORTIA IN AGRIC LTURAL LIBRARIES

- To support needs of user' cooperation, resource sharing and enhance access to electronic information.
- To develop suitable collection development and acquisition policy.
- To make use of the available latest information technology.
- To eliminate the different problems faced by the Librarians in order to provide various services to the users.

The main objective of consortia is to control and reduce the information costs, improve resource sharing, develop a networking information environment and to encourage and facilitate interlibrary communication, education and resource sharing within its diverse multi-type library users,

FACTORS NECESSITATING CONSORTIA IN AGRICULTURAL INFORMATION SYSTEM IN INDIA

The following are the factors which influence the agricultural information system towards consortia:

1. Increase of books and journals prices.

2. Limited library budget.
3. Increased cost in providing services.
4. Exponential growth of information in agriculture.
5. Inadequacy of library materials required for single agricultural library.
6. Increase in variety and degree of user needs.
7. Development of technology.

SHORTCOMINGS OF THE PRESENT SYSTEM

In the present situation, the services of agricultural information system have been found to be far from satisfactory. The drawbacks observed are mainly on account of the following.

1. Non-availability of materials.
2. Lack of sufficient funds.
3. Managerial problems.

NEED FOR AGRICULTURAL INFORMATION CONSORTIA FOR INDIA

The extraordinary growth and dynamics of information gives rise to financial pressure upon agricultural libraries and information centres to coalesce into consortia in order to share the resources rather than duplicating them. Soaring cost of reading materials as against tighter budget is also one of the reasons for consortia of library and information system.

Agricultural scientists of various disciplines who are engaged in the field of research, teaching and extension are finding it difficult to keep themselves abreast of new developments within their areas of specialization. It is recognized that current agricultural information relevant to agricultural scientists is dispersed in a large number of research reports, periodicals, and other documents. Added to these the range of variation in information needs, level of sophistication and capacity for effective application of information among the potential users (farmers) will result in creating crucial and difficult task for the agricultural scientists to have access to timely, reliable, and precise information.

The present environment in India in relation to the availability of agricultural information sources to the various agricultural scientists will make us to think on two aspects of planning on agricultural information system primarily on the accessibility and awareness of agricultural information sources available in various agricultural libraries and research stations of India. Secondly, providing nascent information generated and which is available outside the nation, i.e. providing access to national and international agricultural databases required by the agricultural scientists of India.

ESTABLISHMENT OF MANAGEMENT COMMITTEE

A consortium is to be established in Indian Agricultural Research Institute (National Agricultural Library of India), New Delhi. It acts as a Coordinating Centre for participating libraries. It can be formed with the representatives of ICAR Institution Libraries, State Agricultural University Libraries and Deemed University Libraries. The Management Committee may consist of 15 members with the Director General of Indian Council for Agricultural Research as the Chairperson, and Director of IARI, two Vice-Chancellors from State Agricultural Universities (on rotational basis for two years), five University Librarians from State Agricultural Universities (on rotational basis for two years), three librarians from ICAR institution Libraries on rotational basis (again two years), and a representative from Ministry of Agriculture, Government of India. The Librarian of Indian Agricultural Research Institute will act as Member-Secretary for the consortium.

FUNCTIONS OF CONSORTIUM COORDINATING CENTRE

The Consortium Coordinating Centre would function under the management committee for inter-institutional coordination. The important activities of the consortium coordinating centre would be as follows:

- To act as nodal agency for increasing the cooperation amongst the participating institutions.
- To coordinate all activities concerned with subscription of e-resources on behalf of the consortium.
- To liaison with electronic publishers to provide training and technical help to the participating member institutions.
- To organize the meetings of the Management Committee and to decide upon the policy issues.
- To maintain the Website for the Consortium for the benefit of its members and to encourage sharing of resources in an online mode.
- To organize annual meetings of the consortium members.
- To propagate the Consortium with other institutions and enroll new members in the consortium.

FUNCTIONS OF CONSORTIUM

Every agricultural library differs from one another according to its collections, information needs of users, working methods, source of finance, processing of information, etc. The various steps of functions can be adopted by a consortium for effective functioning by standing on a common platform, which are:

Agreement for Establishment

A concrete agreement needs to be established for the participating libraries in consortium to achieve a common target. All libraries should work in such a manner that they are mutually exclusive, but they have to follow the common mission of a consortium as a whole.

Administration of Library Consortium

To run the consortium smoothly, a statutory body is very much essential to be formed taking Library Heads from every library, the Librarian, Indian Agricultural Research Institute (IARI) as Chief Coordinator. Each library will inform about their collection, databases, user services, and training of human resources for handling new equipment, etc. to the Chief Coordinator.

Financial Control

Due to financial crisis, libraries form a consortium for providing better service to the users. The crucial activities of the members of management committee is to take action on the following financial points like:

- Whether a consortium fund be created to subscribe for the core journals in different subjects in multiple copies at a discount rate.
- How to manage the fund of a consortium and how much money is to be collected from each library as a contribution.
- ICAR shall take the responsibility to negotiate with some publishers for procuring the resources required by the consortium members.
- Standard accounting methods have to be followed for the fund raised by the Consortium.

Joint Work

As per guidelines of a consortium, individual library will prepare a list of title for resource sharing among participating libraries looking into the user demand and may also initiate to contact with different publishers for negotiation. All the participating libraries under library consortia must work jointly just like a joint venture in business sector to make the consortia a total success. Each library will subscribe the core journals and prepare the union list of titles to be covered under resource sharing programme and the member libraries may also send full contents page service of all issues of the title to another library.

Evaluation

After a certain period, the whole activities of a consortium (i.e. individual as well as co-operative) must be evaluated on the basis of the pre-determined objectives. If there is any lacuna in between, the work performed and to be performed should be adjusted by taking correction measures in activities in future. If there is a need for any change in the path of action, it should be changed immediately to achieve the objectives of a consortium.

CONCLUSION

A consortium plays an important role in agricultural libraries and success of consortia totally depends upon the members of consortia. The agricultural libraries should also make use of the available technology to meet the expectations of user information requirements. An agricultural libraries consortium plays a vital role in India providing information to agricultural scientists in right time. It helps to increase production and to achieve self-sufficiency.

REFERENCES

1. Biswas, Bidhan Ch and Dasgupta, Swapan K (2003). "Opportunities for libraries in managing and resource sharing through consortia: A new challenges for Indian libraries". In. Mapping technology on Libraries and people. CALIBER. 2003, 291-97.
2. Kaul, H.K 1999). Library Resource sharing and networks. Virgo publications, New Delhi.
3. Murthy, TAV (2002). Sharing and consortia for India. IIT Seminar, Kharagpur, pp.14- 15.
4. Veeranjaneyulu, K and Rajagopal, T. (2001). Networking of agricultural information centres in Andhra Pradesh (AP-AGRINET) : A model. International Library Movement 23(3), pp. 162-176.

COMMUNITY INFORMATION SERVICES FOR FARMERS

INTRODUCTION

Jawallarlal Nehru, while laying down priorities for national development in post-independent period stated that 'every thing else can wait but not agriculture'. He dreamt of India that would become self sufficient in food requirements first, to give way to self sufficiency in other respective walks of national importance. Ways and means were therefore found out and practiced to ensure considerable increase in agricultural products.

AGRICULTURAL INFORMATION

The boost in agricultural products depends on the following factors:

- The speed at which research in agriculture is conducted.
- The speed at which such results are communicated to the ultimate consumers/users namely formers, youth, and other rural people, and.
- The speed at which experiences gained by the ultimate users of research results arc communicated to the researchers as feedback.

Significance and role of agricultural information for agricultural development was realized after independence when agriculture scenario changed fast and new techniques were being used in agricultural production in Indian agriculture. Farm information was regarded as one of the inputs in agricultural production and farm, information was made available through various sources.

The fruits of research in agriculture for the improvement of productions essentially require timely supply of information to farmers and motivating them to implement the packages and practices developed for each crop. It is not possible to achieve self-sufficiency in food production without adequate information supply to farmers.

INFORMATION FARMERS

Farmers are the ultimate users of the agricultural research findings. Accordingly, such a category in the information communications process is likely to include illiterates as well as the literate strata. Therefore Community Information centres have a significant role to play for communicating new ideas and practices to the farmers.

For progressive educated farmer's best means of communicating research results for the library is through publications seeing his influence and following the example, others in the community likely to adopt the new techniques. While issuing publication for the purpose, the Community information Centre has to pursue the material from different aspects to ensure in a way the ever-lasting impressions on the farmers mind.

The UNESCO manifesto believes that the public library is a living force for education, culture and information, and is an essential agent for fostering of peace and spiritual welfare through the minds of the people.

COMMUNITY INFORMATION SERVICES AND FARMERS

The following are some of the important services expected by the farmers from Community Information Centres

1. Availability of requisites (seed, fertilizer, equipment, pesticides and others).
2. Environment (climate, soils and topography).
3. Market requirements and opportunities.
4. Information regarding Kisan Melas, Rythu Sadassu etc.
5. Information regarding agricultural programmes in T .V., Radio and Newspapers.
6. Information regarding welfare schemes.
7. Information regarding financial facilities like crop loans etc.
8. Information regarding employment.

9. Adult education programmes.
10. Land registration information.
11. Crop insurance schemes.
12. Development of Horticulture and different schemes
13. Health information.
14. Current developments and research in agriculture.
15. Exports & Import policies like WTO etc.
16. Organizing training programmes for farmers.
17. Internet services.

PLANNING COMMUNITY INFORMATION SERVICES FOR FARMERS

Community Information Centres should provide the following services for farmers:

1. Illiteracy, which is the main constraint for proper interaction and active participation of farmers, to over come this problem adult education programmes may be organized by the Community Information Centres.
2. The farmers should read popular magazines, watch T.V. and other Audio Visual aids. The illiterate farmers who will never be attracted by the community information centres can easily be impressed by methods of this kind. The Audio Visual Media not only provides them the required information but it also gives them fun and enjoyment. But our centres lack such communication media. Public libraries need to preserve and make use of audio visual communications particularly for farmers. But our public libraries do not have equipments like Radio and Television. The government organizations and voluntary bodies should come forward to install such equipment in the Community Information Centres.

3. The Community Information Centres should arrange exhibitions, displays, film shows, lectures and adult classes, thus contributing directly to the economic, social and culture development of farmers.
4. The Community Information Centres should arrange cultural activities like drama where illiterate and neo- literate farmers participate.
5. Organization of farmer clubs helps the farmers to create awareness of latest developments in agriculture and to adopt them.
6. The agricultural scientists and extension workers should visit Community Information Centres regularly to know the problems of farmers.
7. Community Information Centres should screen regularly films on the latest developments in agriculture.
8. Agricultural Universities should establish Agricultural Information Centres for fulfilling the information needs.
9. The information and public relation department is the mouthpiece of the government policies and prograammes. It should open information centres in every community and should screen regularly films on the farmers' welfare schemes.
10. The Community Information Centres should display posters relating to agricultural information, welfare schemes, crop insurance schemes, marketing channels etc. It should display the T.V. and Radio programmes relating to farmers well in advance.

Internet services should provide to the fanners. Recognizing the benefits of internet for rural development the government of Andhra Pradesh came forward to provide Internet connection at each mandal so as to reach its use to rural people. It is right time for public library systems to extend Internet services to rural people through rural community information centres or rural libraries.

REFERENCES

1. Mohan, Surendra(1977). Agricultural Information Systems and Services. *Journal of Library and Information Science* 2(3).
2. Raina, Roshan(1983). Libraries in agricultural research communication, *Lucknow Librarian* 15(2).
3. Richardson, D(2000). Internet for rural development. *Kurukshetra,*

INFORMATION SOURCES AND SERVICES FOR RURAL DEVELOPMENT

INTRODUCTION

Free flow of information to the ultimate consumer, has been a subject, of concern to librarians, documentalists, scientists and of late to governments also. Right to education and right to information; of honestly conferred and sincerely enforced could bring about major social changes among rural masses. Right to information is one of the emergency human rights of the twenty first century. If people are well informed, they will be more vigilant and therefore democracy is bound to become more vibrant. Right information to right people at right time will naturally help for the socio-economic development among rural masses.

Food, Clothing and Shelter are the basic and essential needs of the human beings. All the governments in the developing countries want to provide the higher levels of living for their people. This goal can be achieved effectively only through massive programs of directive change, generally known as development programs. Development is a type of social system that produces higher per capita income and improves standards of living through modernized production methods and improved upon the farming. The efficient farming to help the people to overcome poverty: Farming can be done effectively when the proper techniques and methods are used. Farmers must learn and apply improved ways of farming to have growth in their agriculture as well as economic field. Many interacting forces are included in rural development, education for transfer of technology is one of them.

Development starts with the people and their education organization and discipline. Development can be achieved only with the people's acceptance of the change. Generally advanced technology is not reaching the people due to lack of communication. Majority of the people is not having the exposure to the development programs and technology especially rural and weaker section of people.

INFORMATION AND RURAL DEVELOPMENT

Information has become an essential ingredient for the progress of civilization and society for all the times. Lack of information transfer and its effective means of communication are now widely recognized as being limiting factors in socio-economic development of the people. The nation and the individual could survive only with the provision of information governing every single act of man and society. The nature in terms of its accessibility is always linked with the development of Nation.

UNESCO is convened that "Information is an essential part of nation's resources and access to it is one of the basic human rights; information is not only a national resource vital for the scientific and economic progress, but also the medium of social communication". The personal, vocational and social development of the individual depends on the quantity and accessibility of information.

GENERATION OF INFORMATION

Information is generated tremendously at various levels by various groups of people. For example Scientists, technologists, academicians, legislators, leaders etc. Information does not have value until it is used. Various methods and processes are involved in different stages for the optimum use and transfer of information. The information flow needs proper processing for optimum accessibility and usability. The process involves collection, organization, storage, retrieval and use of information.

INFORMATION SERVICES

The totality of the services provided by libraries and other information centres are made up of several different types of services to meet the demands for the provision of Specific documents identity and named by the users; the provision of documents, which the users does not specify or identify individually; and of information drawn up from documents by the staff of the service. The last service, a type of information analysis service, can be of a nature that assumes a demand before it is arises, thereby requiring the staff of such an information service to take a very active role in information gathering and disseminating. We can divide these information services also as primary, secondary and tertiary.

We can say Selection, Production and Distribution as primary services; Analysis, Storage and Dissemination as secondary services and finally Evaluation, Compression and Consolidation as tertiary services.

Functions of some information services in minimizing the difficulties of information users are given below.

1. Current Awareness Service, Extension services, Publications and Publicity provided to minimize the gap between generators of the information and users.
2. Indexing, Classification and Storage systems reference services are provided to select the information from over abundance.
3. Reviews, State-of-the art and trend reports, digests, information analysis and evaluation are helpful to the users to obtain information quickly for specific problem area from accelerated growth of information.
4. Indexing techniques to interlink subjects, information analysis and enquiry services are useful to get information of a restricted subject field from interdisciplinary nature of information scatter.
5. To provide the information available in multiple languages to the user, translation service is essential.
6. To select some standards and patterns from the available wide range of standards and modes of presentation of ideas: selection and presentation or repacking information to users needs is essential.
7. It is difficult and time is inadequacy for evaluating and selecting information, so information and data analysis and evaluation are helpful to overcome the problem.
8. Obtaining copies of documents, reprographic facilities are used to reduce the delay in final stage of document provision.

INFORMATION SOURCES FOR RURAL DEVELOPMENT

1. All India Radio and Doordarshan should broadcast rural development programs regularly.
2. Organization of village clubs helps to know the welfare schemes. It helps to create awareness.
3. As the Village Development Officer becomes mass sources of information about rural development schemes, village level extension networks should be further improved. The Village Development Officer should visit villages regularly to guide the rural people on various development programs.
4. The information and Public Relations Department is mouthpiece of the government policies and programs. They should open information centers in every Mandal and they should screen regularly films on rural development programs.
5. There is an immediate necessity to build up the rural information centers as integral part of the public library system. In developing a network of information transfer to rural and backward areas the public library system shall be considered as a focal point. The system should consist of three tiers viz. 1. District Center 2. Revenue Division Centres 3. Mandal Units with a state level controlling authority at the apex. The state apex Information center should feed necessary information to the public Library System. In the same way the Public Library System should inform the center, the requirements of rural masses.

REFERENCES

1. Calligan, A. (1981). Information and advice services for rural area - review of the country libraries. The City University, Center for Information Science, London.
2. Gangadhara Rao, P. (1986). Rural Information Centers. National Seminar on Assessment of Public Library Development in India.

3. Sambasivan, K and Abdul Rahimam, E. M. (1984). Rural Development and appropriate Information transfer, Alis Bulletin 8 (1 & 2).

4. Veeranjaneyulu, K. (1997). Study of the scheduled castes Awareness about the welfare Schemes. In : Information needs of Scheduled castes: A critical and analytical study.

INTERNET AND ITS SERVICE FOR AGRICULTURAL INFORMATION SYSTEM

INTRODUCTION

The internet and the facilities of world wide web (WWW) offer unique opportunity of near communication. One can access the information for every individual throughout the world. This communication network has created a "Cyberspace" where one is free from the boundaries created on the basis of geography, language, religion or culture. Further development of internet would lead to formation of information super highway, where rapid access to information resources would be possible irrespective of the place of initiation of search on the globe.

ORIGIN OF INTERNET

In 1969, the US department of defence created a network called the ARPA net, (Advanced Research Project Agency). Due of enormous increase in use of ARPA net for non-military purposes, the US department of defence created an exclusively military network called MIL net. A few years later National Science Foundation formed the NSF net, similar but faster then ARPAnet which linked together NSF researches. At that time there were no personal computers. The model of centralized mainframe computers and terminals ruled the computing worlds. Under this model, the big mainframe sat at the center of a starfish-link system with a dumb terminal (a CRT and a keyboard) at the tip of each tentacle.

Internet is network of computer that offers information and access to people. The world "internet" was coined from the word "inter connection" and "networks" such network of networks is called "INTERNET"

ADVANTAGES OF INTERNET BASED SERVICE

The main advantages of internet based services are as follows.

- The librarian can serve even the remote users sitting at his reference desk.
- The users can lanch a reference query through e-mail.
- The users browse the particular site for his need.
- The users time can be saved to a great extent.
- Current and up to date information can be provided.

WEB SEARCH ENGINES

Internet users can have access to number of search engines through which a bulk of information can be collected. Commonly used search engines are :

I. AA-IN-ONE

Home Page address	:	http://WWW.albany.net/allione/.
Search method	:	Key word
Comment	:	Efficient means of using many search tools

II. ALTAVISTA

Home Page address	:	http://W W W.altavista.digital.com
Search method	:	Key word
Comment	:	It has the reputation of having the most sophisticated search system.

III. EXCITES

Home Page address	:	http://W W W. Excite.com
Search method	:	Subject and Key word
Comment	:	Excite search is very good for the beginners.

IV. INFOSEEK

Home Page address	:	http://W W W.infoseek.com
Search method	:	Subject and Key word
Comment	:	New site submissions are added immediately.

V. YAHOO

Home Page address	:	http://W W W.yahoo.com
Search method	:	subject and Key word
Comment	:	Yahoo! Search is very useful for the beginners and it is easy to use.

VI. LYCOS

Home Page address	:	http://W W W.lycos.com
Search method	:	Subject and Key word
Comment	:	New site submissions are added immediately.

VII. MAGELLAN

Home Page address	:	http://W W W. magelan .excite.com
Search method	:	Subject and Key word
Comment	:	New site submissions are added immediately.

VIII. ELECRICMONK

Home Page address	:	http://W W W.electricmonk.com
Search method	:	Subject and Key word
Comment	:	New site submissions are added immediately.

INTERNET SERVICES FOR AGRICULTURE INFORMATION SYSTEM

Today the internet is growing tremendously and is known mainly for the services it provides. Some of the best known agricultural information services available on the internet include the following.

I. E-Mail

In addition to its use as a cheap and fast communication system. It can be used to join many electronic mailing lists related to Agriculture and allied subjects.

As soon as one sends an e-mail the other will receive an automatic reply with information about registration, methods of accessing past files. From then on, one receives all messages posted to the mailing list.

II. Journals Websites

It is now possible, through the internet to browse many foreign journals on the same day of its publication without even subscribing to it. Many journals like, journal of Agribusiness, Journal of Extension and Progressive Farmer are available on the internet.

Simultaneous release of contents, abstracts, and selected full paper on the internet as they publish, is an important use of the internet. Instructions to the authors are also available for many journal in their web site URL of some journals are given below:

Title of the journal	URL
Journal of Extension	http://WWW.joe.org/
Progressive Farmer	http://WWW.progressivefarmer.com
Journal of Agribusiness	http://WWW.agecon.uga.edu/job
Agriculture & Applied Economics	http://WWW.agecon.uge.edu/jaae

III. Association/Conference News

A number of agricultural associations/societies have established their own Websites. One can learn about their activities, conferences etc., by visiting their site. The URL of a few associations and agricultural websites are given below:

Description	URL
Agri Launch	http://WWW.agriaunch.com/
CAB international	http://WWW.CABinternational.com
Agricultural News	http://WWW.@griculture on line.com
The library of Congress	http://WWW.Icweb.loc.gov/
The Action Aid India	http://WWW.actionaidindis.org/
Consultative Group on International	http://WWW.cgiar.org/
Trees and people program & network	http://WWW.trees.shu.se/
Food & Agricultural Research	http://WWW.fao.org/waicent
Indian council of Agricultural Research	http://WWW.nic.in/icar/
Kisan.net	http://WWW.kisan.net/
National Academy of Agricultural Science	http://WWW.naas.edu/
Wagenlngen Agricultural University	http://WWW.wau.nl/

IV. Agricultural teaching on the Internet

The web is an enormous teaching resource for undergraduate and postgraduate agricultural students and also to update most of the agricultural information. Most of the prominent teachers have lecture notes on the web, which can be accessed from any place in the world.

Description	URL
Agri-info	http://WWW.aari.infolink.com
Kisan.net	http://WWW.aari.kisan.net/
Lampman's agricultural page	http://WWW.lampman.skea/cap/agriculture

V. Jobs/Fellowships

The details of job vacancy/Fellowships are available on the internet at the appointment columns of journals. Some of the Website are given below :

Description	URL
Virtual Farmer show	http://WWW.farmershow.net/chet/
Career Opportunities	http://WWW.agricareers.com
Agrisurf	http://WWW.arisurf.com

CONCLUSION

The internet can bring enormous future benefits to agriculture information system. Internet offers numerous possibilities for keeping abreast of new development. Though anybody with internet can access and avail the services on the internet, since users have insufficient time and knowledge of resources available on the internet. Information scientists/librarians can offer many new services using these services on the Internet.

TECHNOLOGICAL AND OTHER ASPECTS OF AGRICULTURAL INFORMATION TRANSFER IN A DIGITAL ENVIRONMENT

INTRODUCTION

Information in digital libraries is electronically stored and accessed. Access to the digital library is, therefore, not bound by space or time. These libraries provide access to information via electronic gateways to remote digital and digitized databases. The access is available to well established, wide range of internet-accessible sources of information. One can enter the world of three-dimensions, created by computer imaging and also interact with the surroundings imagery and get immersed in it, with appropriate responses from the "virtual environment". To cope up with the information needs with speed and relative accuracy and reliability, the digital library is the most important and reliable resort, and so knowledge discovery in these type of libraries becomes a predominant fact.

With the advent of information society, libraries, being part of the society, are gradually bringing in the speculated changes in their structure, function and services. Traditional libraries are shifting their budgets to electronic services, CD-ROMs, online access etc. The libraries are gradually getting connected to networks, nationally and globally. The existing library collection has shifted from micro document to micro piece of information. Information storage has also changed its medium - from paper to Electronic. In addition to the books and journals, collections now range from CD-ROM data bases & Multimedia databases. A time is to arrive in the very near future when the concept of "Virtual library", (i.e. a library without walls but providing access to resources) will not be merely a concept, but a reality. So the process of development is already in; the traditional libraries shifting towards automation, automated ones towards electronic; then in the further movement, these electronic ones moving to digitize their collections and ultimately a virtual library is the end mark of the process.

WHY SHOULD WE GO FOR DIGITAL SERVICES?

The digital environment has had a profound impact on the form and format of information services. The emphasis is on the network centric services. That is to say that the information items are retrieved and made available to the end user at the work place itself. Again each user and instance of the use of the system may be unique. In a case, it becomes very difficult to design a reasonably acceptable model across the sections of the users.

The first consideration is whether the information itself or the information regarding the item will be delivered. The traditional forms of bibliographic services, of course, have their significance. But in an online distributed system, such secondary information services should also give a real time linking to the documents. Copyrighted information is not readily available to be offered to the end user in networked environments. This is a major hurdle for digital library projects that aim to digitize and store collections online.

The second consideration is the time factor. As emphasized earlier, the need of the moment is the information itself less often the items containing the information. The user does not have the time to go through a larger retrieved set to locate the appropriate information. In such a case, the responsibility of analyzing the information needs and supplying the pertinent information lies upon the digital library. The Information maybe in one or more documents and in more forms. For an effective information service, considerable efforts at information consolidation and presentation have to be made. Again time and manpower are important factors.

The Third factor to be considered is the portability and flexibility of the output. Once a user retrieved, it should be in form and format that is acceptable by most systems in use and must be easily adaptable to other systems. Standardization of the bibliographic information of data interchange is much beaten track. In any case it has a special significance in the digital library environment as the retrieved information may often be used as a link or input to some other system.

NEED FOR AGRICULTURAL INFORMATION TRANSFER

In the Agricultural sciences, transfer of information is a fundamental and essential prerequisite to the improvement of agricultural products, techniques, yields etc. Ultimately, the only valid purpose of agricultural information work is to help farmers more food, and more nutritious food, thus making it easier for the individual consumer to get the food he needs at a price he can afford, and help to relieve mankind of the threat of famine.

The ultimate goal may be addressed directly, by providing needed information *to* farmers in a form that is useful to them and in a language they can understand. Or the goal may be addressed indirectly, by providing information to extension workers, who will interpret it for farmers research workers, who will apply it in their efforts to develop more effective agricultural, practices and to teachers, who will use the information to improve their training of a future generation or agricultural workers.

The recognition will come when we can show what we have done to improve agricultural production, and when the other groups within the agricultural community tell the authorities that and services that we provide.

The technology already exists, the equipment required is relatively cheap, the obstacles to development are essentially the inertia of the authorities that regulate such activities and that set tariffs.

TECHNOLOGICAL AND OTHER ASPECTS OF AGRICULTURAL INFORMATION TRANSFER

Individual producers are facing pressures to adjust rapidly their production and management if they are to survive in economies that are dominated by the manufacturing and territory sectors. The costs of labour, fuel, fertilizer and other farming inputs are generally determined by broad social and economic policy. At the same time, changes in the world economy, agricultural policy in other countries and fluctuating international relations can have a dramatic effect on demand and prices for agricultural products.

Within this context the individual producer requires information on prices and outlook and new developments in agricultural technology that might improve productivity at the farm level. It is these information needs of individual producers that should ultimately dominate and drive our agricultural information services.

USER REQUIREMENTS IN DIGITAL ENVIRONMENT

The parameters which define user requirements are, in order of importance:

1. Speed
2. Relevance
3. Comprehensiveness
4. Volume
5. Cost
6. Presentation
7. Timeliness
8. Comprehensibility

THE BENEFICIAL EFFECTS OF DIGITAL SERVICES ARE

(a) allows greater sharing of information than ever before

(b) enables correlations, trends to be identified quickly and accurately

(c) improves means of rapid alerting to developments

(d) eliminates some uninteresting, repetitive work

(e) reduces unit costs

(f) positively enriches certain types of work

(g) facilitates processing huge amounts of data from many different sources

(h) increases users awareness of importance of information resource

(i) increases possibility of decentralisation networks

(j) assists effective presentation of information and data

(k) enables lay end user to interact directly with complex systems

(l) destroys the concept of distance

(m) facilitates processing of different types of information-text numbers, pictures sounds

(n) accelerates convergence between different functions.

ROLE OF LIBRARIAN

Undoubtedly the role of the Librarian and Documentalist will be diminished unless he too can participate in dialogues with other members of the agricultural community. The Librarian who admits that he does not have the scientific knowledge to do subject indexing is also admitting that he does not have the scientific knowledge to have effective dialogue with his clients. The client may recognize that such a Librarian is the custodian of a store of relevant knowledge, but he will not trust that Librarian to find the appropriate items in the store. How can an existing librarian without an agricultural background upgrade his knowledge so that he can effectively communicate with his clients.

Many agricultural Librarians and documetalists are located in institutions that also have teaching functions. With advice from the faculty, we can find first the more introductory courses and then the more specialised courses that will add to our understanding of the subject and needs of our clients. Once one has determined to acquire more knowledge in the subject area, it is amazing how much can be achieved even if the technicalities are initially some what frightening.

Agricultural Librarians and Information workers in industrialised countries are facing new roles and new challenges in contributing to the development of these new services and development of these new services and in relating to new groups of users.

CONCLUSION

The wide range of information needs of different groups in the agricultural sector presents a number of challenges in providing timely and relevant information services. Library and Information workers are finding opportunities to apply their skills in identifying, organising, classifying and selecting material in both conventional and non conventional formats to meet the changing needs of their users.

In many industrialised countries new communication technologies are starting to have a major impact on the provision of this information. Videotext, computer networks, telephone dial up services, television and radio are all being used to facilitate the flow of information of different User groups.

REFERENCES

1. Thorpe, P (1980). Agricultural information services for the third world: Problems, developments and prospects. Quarterly Bull. of IAALD 25(2/3), pp. 29-37.
2. Moll, P. (1982). IT and Third world - aid or aggravation. Focus 13(3), pp. 26-29.
3. Woodward, A.M. (1980). Future information requirements of the Third world. Journal of Information Science 1, pp. 259-265.
4. Woolston, J.E. (1983). The future of Agricultural Information Transfer: Human and Institutional Aspects: Paper presented at the International conference on Education and Training for Agricultural Library and Information Work, Nairobi, Kenya, pp. 34-39.

QUALITY MANAGEMENT IN LIBRARIES AND INFORMATION CENTERS

INTRODUCTION

Of late there has been a lot of talk about improvement of quality in the manufacturing industry. Quality has become imperative if the local industries are to survive in an open economy. Moreover, in order to be able to export to the European countries, a company must have the ISO 9001 accreditation, which by itself requires a high standard of quality. However, unfortunately there has been no mention of quality management in the service sector in general and in the libraries and information centers in particular.

Libraries constitute an important segment of the service sector where the concept of quality can be applied profitably. Quality being subjective is difficult to define. Broadly speaking it represents the totality of features and characteristics of a product or service that bear on its ability to satisfy stated or implied needs. This brings us to the question of what are the stated and implied needs. These ideas will be given better shape as we proceed along.

Application of the simple but crucial concept of quality will entail untold benefits for the users. However this requires the active cooperation of the library management, library professionals and the users.

THE LIBRARY MANAGEMENT

The role of management in enhancing quality need not be over-emphasised. It has a say from the day it has been decided to set up the library. The locality and the building in which it is housed are to some extent a reflection of the management's view of quality, If it is to be recognized as a quality library with respect to its services, then the management should be prepared for bold investments in terms of money, time and labour. A brand name cannot be achieved without investments.

This is not to say that management has no role in enhancing the quality standards in an old library. Adequate ventilation, lighting and furniture are some of the necessities of a good library. However, the management should evolve administrative and financial procedures which are conducive for libraries to function in a professional way. Scientific principles of management are to be followed to run the library for attaining quality standards.

More importantly, the management should release funds for the purchase of books and periodicals regularly. Technical committees should be formed for different subjects. These committees should have the responsibility of buying multiple copies, even then quality service can be rendered to the users by some imaginative planning. In such cases, two books can be bought and only one of them can be lent out of the library.

THE LIBRARY PROFESSIONALS

Quality cannot be infused into this sector without the active cooperation of the library staff. In this fast changing computer world, these professionals should be forward looking if they are to be a part of the quality movement. They must constantly update their knowledge. For this purpose, they should regularly attend technical workshops and advanced short term courses. Interaction with the users is equally important. It is only then possible that they will be able to understand the real needs of the users better.

The sense of service to users need not be over-emphasized. Perhaps it is one of the principal ethics of librarianship more so if it is so be of a quality standard. The library staff must be able to feel the needs of the users. They should endeavor to keep the books and journals at the proper shelf. The concept of quality demands that books and records must be properly maintained. All the catalogs – author, title and subject must be updated with the addition of new books. In case of joint author, the book should be traceable at anyone of the authors. Moreover, quality demands that records be reproducible. If a reader wants to know as to who has taken a book and when it is due, the library staff should be able to produce it correctly.

THE CLIENTELE

At present, the clientele are not very demanding. But with the passage of time thing will change. The clientele can influence the quality standards of a library by constructive criticisms and suggestions. Perhaps they can also exert influence on the management for provision of better facilities. The professional development of the library staff can occur only as a response to the exacting and challenging demands of the user.

IMAGINATIVE PLANNING

In order, to provide better quality service at least one reference copy of important and costly books should be maintained. This will enable to user to go through and satisfy his need to a large extent. If his requirements are urgent, he can take xerox copies of limited portions. This calls for the installation of xerox machines in the library. They should be suitably charged so as not to incur loss.

Making a motto to render quality service, implicitly assumes that the user is the king. And hence he is to be kept happy. Hence library rules should have a provision to reserve books if presently the book in question has been issued out. Of course, after a certain period of time - say a week - the book should be replaced back at the shelf. Mention should also be made of inter-library loans. If a user cannot get the book in the library, then the library personnel should be able to procure it for him/her from other lending libraries in the neighborhood. This will only be possible if the libraries maintain a network among themselves.

STANDARDIZATION

Naturally, if every library is to endeavor to attain quality standards, then there should be an apex body which can give appropriate guidelines. This, calls for the setting up of a nodal centre for coordination among the libraries. This centre should have the infrastructure for carrying out research work in the areas of indexing, cataloguing, among others.

Specific guidelines should be issued with respect to these important facilities.

Moreover, the nodal centre should also aim for standardization in the library environs with respect to the more mundane things. Clear guidelines regarding quality should be issued with respect to the type of the catalogue cards, furniture, cabinets, lighting conditions etc.

Dr. S.R. RANGANATHAN'S FIVE LAWS IN QUALITY IMPROVEMENT

Dr. S.R. Ranganathan, venerated as 'Father of Library Science in India has formulated the fundamental principles of Library Science on the basis of his own experiences, observations and experiments on the working of the libraries. He enunciating the famous five laws of library science in 1931.

The meaning and significance of these laws with reference to quality improvement in libraries and information centers are as fresh and relevant today as when they were enunciated in the thirties. These laws, which are classical, have been interpreted, evaluated, examined and tested in India and abroad by many library scientists.

LONG TERM PLANNING

For improving, the conditions of the more important libraries a long term planning perspective is necessary. A network of important libraries in a region should be developed. This will greatly help the users in seeking reference works. Inter-library loans of books and journals (original or in the form of xerox copies) should be more encouraged. Indexing of important subjects which hitherto has been greatly neglected must be immediately taken up.

The role of computer in libraries needs no introduction. They provide ease of access and centralized access to an increasing variety of networks of information. More importantly they allow a user to store edit and analyze information more easily, systematically and quickly. These suggestions require bold investment in terms of money, time and labour. Perhaps, they also call for a change in the curricula of the library sciences followed in most universities.

CONCLUSION

Quality in the library is perhaps the sum total of the commitment to quality of the management, staff and users. The image of a library will be in direct proportion to the quality of service which it offers. In today's fast changing scenario, the intense hunt for information will increase many fold in the near future. The clientele will increase and will demand quality in the service offered. A library which fails to respond to the new challenges thrown up by the ravages of time will loss its name, fame and clientele. It will stand as a mute testimony to the world that those who do not keep pace with time and quality have no place in today's fast moving computer world.

REFERENCES

1. Ton, Carbo Bearman (1984). The changing role of the information professional. Library Trend (Winter).
2. Elizabeth, Futus (1987). What constitutes a "Good" collection?". Library Journal.
3. Mittal, R.C. (1984). Library Administration theory, and practice. New Delhi: Metropolitan Book Co.

DIGITAL LIBRARIES: THE PILLARS OF QUALITY ASSURANCE

INTRODUCTION

Digital medium is fast replacing the paper in a dramatic revolution. This development is attributed to several reasons. The traditional library methods of information transfer are incapable of handling large data, and servicing users effectively, especially in resource sharing environment. It is also attributed to electronic databases, for the amount of data that they store in mind boggling. It is also attributed to digital technology for its ability to handle text and other media elements such as sound, graphics, pictures, colours etc.

QUALITY OF INFORMATION IN DIGITISED FORMAT

The latest technologies offer cheap computer processing power, cheap mass storage, inexpensive and ubiquitous access to high speed networks, and retrieval devices. These give us the ability to create, manipulate, store and transfer large quantities of information in digital form at low costs. Though conversion from the conventional storage to digitised format requires deft handling and discrete analysis, the information stored in digital form can be had as hyper books on CD-ROM.

Hyper Books

Hyper books are electronic books defined on the basis of the paper back metaphor. Metaphors have been widely used to facilitate learning of computer system. It is a model with which the user is familiar and it enables users to exploit their existing knowledge of other domain when learning to use a system.

Hyper book contains text and graphical images. Graphical objects are always visible and appear in the text much the same way as in a paper book. The feature of the hyper text is that it contains embedded links i. *e.* active text that is connected to other parts of the book.

CD-ROM

The CD-ROM stores multimedia data permanently .The end user cannot put his own information on it. The essential features of CD-ROM technology are as follows:

- high storage capacity
- no head crash low error rate, correctable data
- long life
- random access
- easily transportable
- OS independent file system
- its multimedia, multiplatform and multi language phenomena.

DIGITAL RETRIEVAL SYSTEM

The Internet: It is the world's first digital information utility, an all purpose network that lets anybody send anything digital to anyone anywhere. Without argument, the Internet is the popular network ever evolved. It knows no geographical boundaries. All its growth has come about spontaneously.

The Internet is a cooperative computer network of networks that links libraries, schools. governments, cooperation's, Individuals and the like to each other and offers vast information resources. Through Internet we can send messages and information in file format to any part of the world by sitting at one place. This of course is a wonderful aspect of digital information system.

QUALITY ASSURANCE THROUGH DIGITAL LIBRARIES

Organizations are increasingly replacing conventional storage and network system with digital systems. When we compare the modern technology based systems with traditional systems it is found that the library personnel can render better quality services through digital libraries, this is because there is least burden, and the services are more attractive. Further, we can make use of digital systems in reasonable time period. As Dr S.R. Ranganathan states in his fifth law library Science, "Save the time of the reader". We can do it through digital libraries. Regarding quality, it is always a supplement in timely rendered services. The quality of any document is further assured through its long life. More and more information can be stored in a little space in digital media, thereby it solves the space problems of libraries too. Quality of information in large quantities can be obtained in a quickest possible time.

ROLE OF LIBRARIAN IN DIGITAL ERA

Our mission is to work towards meeting the information needs of people. These can be done by providing higher quality information systems which utilise current and emerging information technologies i.e. digital technology.

Libraries and information professionals who are not willing to learn the modern technological advances must worry to justify their existence. The basic concept of quality management has to be applied library and information units for their existence and success. Responsibility, performance and control are as essential for the successful performance of libraries and information service agencies as they to any other enterprise.

Technology alone can not bring about the required change. The attitudes, practices and policies need to be changed also, if the libraries in India are to benefit themselves and their community of users by application of new technologies, digital information environment is going to be part of our daily professional activities. In the next couple of years, simply the way we carry out research, teaching and learning is not going to be the same at all.

Since computers do cataloguing now, the librarians who navigate the Internet will only be able to survive. Others will become part of the history. If you want to be called 'cybernarians' learn to dabble with cyberspace technology. The exchange structure of the digital data types lets you leverage on the information content in the ways you have never imagined before. But exchanging data across platforms requires lots of hard work.

Irrespective of whether we are working in the public, academic or research libraries, multimedia interaction, digital communication and visualization need to be studied and mastered. This will help professionals to play role in the emerging areas on the information super highway.

CONCLUSION

It is clear from the above, that quality services can be achieved through digital libraries. Developments in the application of computers in publishing, in e-mail and the paperless office, in computer conferencing and other manifestation of interactive television will allow us to conceive a paperless information system of the future.

Further, the digital libraries serve as pillars of quality assurance as they assure us better services in a speedy way and a longer life of the document is also assured for future generations.

REFERENCES

1. Parekh, Harsha (1962). Social Science Literature: Use, Pattern and Biographic Services. *Economic and Political Weeks.*
2. Sharma, S.K (1995). Computer Networks for Library Development in Preparing Libraries for the 21st Century.
3. Catenazzi Nadu De (1994). Hyperbook is a Formal Model for Electronic Books. *Journal of Documentation.*
4. Keys, Marshall. Librarians and Emerging Digital Revolution. *Research Sharing and Information Networks,* 10 (2).

5. Rajasekhar, T.B (1995). Communications of the ACM, *Special Issue on Digital Libraries,* p. 2.

6. Randall, Crunk (1993). Unlocking Data Content. *Byte,* p. 111.

7. Rotheberg, Jeff (1995). Ensuring the Longevity of Digital Documents sci*entific American.*

LIBRARY AND INFORMATION SCIENCE EDUCATION TO MEET THE CHANGING ENVIRONMENT AND MANPOWER REQUIREMENTS IN MODERN AGRICULTURAL LIBRARIES

The progress of research in agricultural fields and improvement of production, essentially require timely supply of information to agricultural scientists. It is not possible to achieve self-sufficiency in food production without adequate information supply to agricultural scientists.

The agricultural libraries are acclimatized to computer environment in library routines as well as information storage and retrieval. Computerization has reduced the pressure of workload in libraries to great extent. It offers efficient services and opens a new era in bibliographic control. It also provides access to required databases in the country and abroad. The computer is capable of storing information for longer period of time in minimum space in electronic form. This information is capable of being recalled with ease and utmost speed, as and when warranted. Achieving progress of research in agricultural sector and proper improvement in production is mainly dependent on the supply of information to agricultural scientists.

IMPACT OF MODERN TECHNOLOGIES ON AGRICULTURAL LIBRARY SERVICES

There are several means of modern technology, which brought in many services to agricultural libraries so as to speed up their activities; they include

- Telecommunication technology
- CD-ROM technology
- On-line retrieval services

- Library networks
- E-Mail
- Internet

New technologies now available will help to remove barriers of distance, of time and drudgery of repeated manual effects in library routines. The advancement in technology will continue to improve upon the effectiveness of the libraries and indispensable for handling messages between libraries and library patrons.

NEED FOR MANPOWER DEVELOPMENT IN CHANGING TECHNOLOGY ENVIRONMENT IN AGRICULTURAL LIBRARIES

The application of modern IT allow to access and store information quickly. The application of modern technology has gradually changed the agricultural libraries. The information technology facilitates a wide range of services, of which libraries could avail themselves and offer to supplement the existing ones.

Some of the areas to which information technology could be applied in day-to-day activities of the libraries are: acquisition, cataloguing, circulation, indexing, abstracting of articles and digitization etc. Therefore it is necessary to impart latest information technology skills and techniques to agricultural libraries to acquire such latest practical skills along with their professional knowledge, it is very difficult for them to cope with the situation in the fast, changing global scenario and at times they may have to loose their ground for survival also. Thus taking the present day situation into cognizance, agricultural library professionals should equip themselves with the new skills and practical experience in the application of modern technologies in the field.

NEED FOR TEACHING MODERN TECHNOLOGIES IN LIBRARY AND INFORMATION SCIENCE EDUCATION

There is a considerable increase in the use of new technologies in agricultural library and information field. The profession is now under the pressure of technological changes. The technological changes need be given an immediate attention. Change such as, on-line information services, change in attitude of Government, user-friendly systems, integration of information activities, rapid introduction of micro-computers and satellite communication are putting pressure on the profession to set new emphasis and direction in Library and Information Science Education. We must adapt with these changes, otherwise we may not be in a position to exit gracefully in the reality of information society. Hence there is need for restructuring of LIS Curriculum and specialization among LIS teachers. Apart from the use of newer technologies, there is also a need to teach fundamental subjects like communications, Hardware, Software, Digitization, networks, CD-ROM technology etc.

NEED FOR IN-SERVICE TRAINING FOR AGRICULTURAL LIBRARIANS

The modem technologies are of a highly technical nature and librarians trained in the conventional library schools are not trained in the know-how of the latest technology and its application to agricultural libraries and information centers. Hence in addition to formal library education, there is an urgent need to provide in-service training for agricultural librarians in the new concepts, skills, and technologies in the fast developing library profession.

CURRICULUM FOR IN-SERVICE TRAINING PROGRAMME

1. Latest developments in Agricultural Information System in India and Abroad.

2. Agricultural Information Processing and Retrieval.
3. Agricultural Information Resource Sharing and Networking
4. Planning for Automation of Agricultural Library
5. Economic Models for Indian Agricultural Libraries
6. Planning for LAN and WAN for Agricultural Libraries
7. Web designing and hosting
8. Impact of Internet on Agricultural Library and Information Services.
9. Information Sources & Gateway Portals in Agricultural Sector.
10. Data Mining Techniques.
11. Electronic Theses and Dissertations (ETD) for Indian Agricultural University libraries.
12. Digital libraries of Agricultural Sector and Building Digital Libraries.

NEED FOR ORIENTATION ON THE USE OF MODERN TECHNOLOGIES FOR AGRICULTURAL SCIENTISTS

As every one knows agriculture is the backbone of Indian economy. Agricultural Scientists seek the help of Agricultural Information Centres for keeping abreast of latest developments. We have observed the networks and research systems developed with the advent of Information technology. But the question arises whether the Agricultural Scientists are making the best use of the networks and systems or not. Sometimes the Research Scholars may be busy in developing new varieties and hybrid and they may not be in a position to look into the latest technologies. Therefore orientation programmes may be conducted by the Library/Information Centre attached to the organization so that the Agricultural Scientists may make best use of newer technologies to keep in pace with the developed notions. The orientation programmes may be short/long term depending upon to boost the morale of Agricultural Scientists but also help us in building the notions economy.

SUGGESTIONS

1. In-service training programmes should be conducted for staff to know the latest information technology.
2. ICAR should provide necessary financial support for Networking and Resource sharing of ICAR Institutes and State Agricultural Universities.
3. User orientation programmes should be conducted for agricultural scientists and students to use modern technologies.
4. It is necessary that courses conducted by library and information science schools should constantly be revised to match the present situation and requirements.
5. Specialized and short-term courses shall have to be conducted to agricultural library professionals to update knowledge. Participation in formal continuing education programs like workshops, refresher courses etc., offers a means for learning new technologies and skills.

REFERENCES

1. Kaula, P. N. (1988). Need for new dimensions to library and information science education in India. *Indian Journal* of *Information, Library and Society* 1 (1-2), pp.58-68.
2. Mangla, P. B. (1979). Manpower development for information handling. *Journal* of *Library and Information Science* 4, pp. 27-32.
3. Rajesh, Singh (1995). Restructuring library and information science curriculum. *University News* 36(38), pp. 8-11.

NEED FOR RESTRUCTURING OF LIS CURRICULUM IN THE PRESENT CHANGING CONTEXT

INTRODUCTION

The history of Library and information Science Education in India dates back to 1911 when Mr. W A Bordon initiated training in librarianship at Baroda. Mr. Asha Don Dickinson also established a training course in librarianship in 1915. Gradually library associations and universities in the country started setting up library schools. There were only 3 universities offering courses, in Library Science in India at the time of our Independence. Today, more than 100 universities are offering courses from B.L.I.Sc. to Ph.D. in Library and Information Science. In addition to that Open Universities are also offering LIS courses.

Library and information Science (LIS) education is a professional education and it is a specialized area of human training relating to management of Libraries. Preparation of librarianship and information work requires a mastery of the body of knowledge and techniques utilised in Library and information operations and services which constitute the discipline 'Library and Information Science'.

The primary aim of education for Library and Information Science should be the providing of the students with knowledge, skills and attitude which will help them to enhance the prestige of the profession.

The library and information Science in India is changing rapidly. Computer culture is already entered in our libraries. A good number of libraries have started using computers for in-house operations and office administration. Online access to international databases is a common phenomenon in many R&D organizations. The electronic publishing, e-Resources and online services are opened new vistas for publishing industry. To meet the challenges and to develop manpower with special competencies the library and information science curricula have to be drastically revised. It should be the objective of a higher degree programme in Library and Information Science to train graduates who are capable of undertaking various tasks of creating information systems and services to meet a wide range of user needs with high quality.

The LIS curriculum should link the education and placement opportunities. The school of library and information science is therefore expected to educate the library and information personnel.

- As a *Scholar* with love for books and other documents, and the ideas contained in them.
- As a *Sociologist* with the understanding of pros and cons of society to whom he or she has to serve.
- As *Scientist* with a special knowledge of Semantics, Logic, Psychology, Mathematics, Statistics and work as a researcher for further education and development of the subject.
- As a *Technologist* to handle everything from the miniature punch card to the Gigantic Computer and its application.
- As an *Administrator* with a special knowledge of scientific management, all in the limited time at its disposal.

NEED FOR TEACHING NEW TECHNICAL SUBJECTS

Till 1960's and early 1970's, library schools aimed at training out library personnel who can run a library on traditional lines. But there was a considerable increase in the use of new technologies in library and information field. The profession is now under the pressure of technological changes. The technological changes need be given an immediate attention. Changes such as, on-line information services, change in attitude of Government, user-friendly systems, integration of information activities, rapid introduction of micro-computers and satellite communication are putting pressure on the profession to set new emphasis and direction in Library and Information science education. We must adapt with these changes, otherwise we may not be in a position to exist gracefully in the reality of information society. Hence there is need for restructuring of LIS curriculum and specialization among LIS teachers. Apart from the use of newer technologies, there is also a need to teach fundamental subjects like Information Communication Technology, Internet, e-Resources, Softwares, Open Accesses Resources, etc.

Library is a real source for the growth of other disciplines. By achieving specialization in the areas, the information professionals will improve the status of the profession and individuals.

TECHNOLOGICAL CHANGES REQUIRED

Rapid advances in information Technology and growing importance of information in daily life have created great demand for professional who can handle information. The curriculum recommended by the Review Committee was apt according to the period of time but because of information explosion, the LIS education transcending its barrier and becoming inter-disciplinary and multidisciplinary demands expertise to handle the amazing data through proper storage, retrieval and dissemination of right information to right user at right time. Hence, LIS Schools must act fast design and implement curriculum, which should prepare graduates to cope up effectively with each emerging situation.

Now, the situation calls for a combination of a wider set of skills with traditional subject knowledge and professional competence with a broader base of general managerial abilities and organization competency IT skills and stronger communication and inter personal skills.

DRAWBACKS/PROBLEMS

1. Even if it is strongly felt to introduce new elements like information technology in Library Science curriculum, there is an insurmountable problem of finding the competent teachers to handle different specializations. But at present our universities are not in a position to create certain posts necessary for these specializations, due to certain problems mostly because of lack of resources and lack of specialist faculty in information technology (IT).

There are only few schools, which have full-time faculty. Some schools are headed by University librarians. Some library schools are even headed by a person of other discipline. This is because of regrettable fact that the negligence of university authorities towards the discipline.

2. In our present curriculum, the emphasis is mainly on theory rather than on practical plane. Most of the LIS schools do not posses the computers of their own to provide computer skills. This prevents the schools in providing training in a real-work situation.

3. This change in curricula will also raise the problem of continuing education for existing faculty members in LIS schools. The retraining will also be needed for the existing staff working in libraries and information centres.

Hence, planning for education and training in Library and information science should be given utmost attention, so that high quality professional at all levels in adequate number and with a variety of skills are consistently made available.

SUGGESTIONS

1. Library schools should make an objective approach to the problem of restructuring of LIS curriculum so as to make its educational and training programs really meaningful in the context of present day needs of the services of information centres in this country.

2. It is not possible to prepare specialists with the help of one year courses such as B.L.I.Sc., M.L.I.Sc etc. We will have to extend the duration of the courses. It may be suggested to introduce library science subject at graduate level throughout India and different specialised areas at M.L.I.Sc. level.

3. Inter-discipline collaboration may be considered. Persons associated with fields like Computer Science, Statistics, Business Management, Journalism and Mass Communications, Information Technology, Printing Technology etc., may be encouraged to take part in the teaching programme.

4. The teacher-student ratio recommended by the UGC review committee be maintained to achieve effective and efficient teaching.

5. The teaching program must be well planned. About one half of the time-table should be assigned to class room lectures and one half to class practical and seminars. Class room lecture could be made much more effective by supplement methods such as demonstration, dialogue, discussion, study groups and talks. A variety of other teaching-learning styles such as individual tutoring, counseling, Laboratory method, Microteaching, simulation techniques should be used for effectiveness.

   But, unfortunately, most of the schools today are not in a position to stick to these norms due to various impediments. Chief among them are shortage of faculty and resources, large number of holidays, loss of working days due to strikes and many others.

6. Specialized and short term courses shall have to be conducted to update the knowledge of professionals working in Libraries and Information centres and also faculty numbers in LIS School. Continuing education of faculty as well as professionals is becoming an increasingly important concern in Indian Library Schools. Participation in Formal continuing education programs like workshops, refresher courses etc., offers a means for learning new technologies and skills like computing and programming skills like ICT, reading, writing and research provide the faculty with better opportunities and scope for keeping up-to-date. This deeper knowledge has to be imparted to professionals.

7. The curriculum must deal in all the components of the Information handling and transfer cycle. It should reflect and remain responsive to changes taking place in society. The views of the employers, students, professionals and educators should be incorporated in the curriculum.

8. The UGC, ICSSR, DST and INFLIBNET should provide more funds for organizing seminars, workshops and other training programs. UGC should organize special training course in IT to

teachers of Library and Information Science as a means of continuing educational program.

9. The U.G.C. should provide adequate grants to the Library and Information Science schools for Micro Computers and for their maintenance.

10. It is suggested that a survey be made to assess the employment potentialities and the manpower requirements in different libraries, information centres and documentation centres, so that specialization in higher level program be planned according to the requirements.

11. New elements like Library automation with special reference to different library software's i.e. SOUL, LIBSYS etc., database administration, networking and resource sharing, telecommunications operations research and systems analysis, Desktop publishing, Expert systems like artificial intelligence, cybernetics etc., economics and marketing of information are some of the areas where specialization is needed. These should form an integral part of our curriculum in order to coap - up. with the emerging trends in Information Technology.

12. It is necessary that courses conducted by Library and Information science schools should constantly be revised to match the present situation and requirements.

13. There should be a provision for lifelong continuing education to meet the real needs of users more economically and effectively and to provide better information services to all the segments of society.

14. The restructured curricula will need further evaluation and improvement with the newer developments that will take place in information and communication field in the years to come.

15. Fully automated libraries should initiate certain diploma courses to inservice professional for coincidence with the professional of the future. In other words they can go hand in hand with newly recruited professionals.

CONCLUSION

A considerable time has elapsed since the inception of LIS education but still many universities imparting these coursers are not properly equipped with infrastructure, faculty and various other indispensable facilities and thus producing inept manpower. So the UGC should take the initiative in curriculum development according to contemporary needs producing a new breed of professional that are able to perform successfully their roles in the fast changing society using changing technologies.

However, it is an admissible fact that many students leaving LIS Schools will inevitably move into libraries where the use of on-line databases, Internet, e-Resources, Softtwares etc., would be lacking. But LIS education is required to provide the new entrants with latest technological skills which will help them to meet the future challenges in the profession. The new technology will supplement the old system and more jobs are created for those information personnel who have computer skills for processing the available information.

Hence, it is necessary to restructure our syllabus of LIS course to keep pace with the development in modem technology. Our schools should train the candidates in different faces of Information Technology to face the challenges of the complex needs of the users.

REFERENCES

1. Amritpal Kaur (2001). Reorienting of LIS education according to contemporary needs. In 46th All India Library Conference proceedings. pp. 520-522.
2. Buckland, Michael K. (1986). Education for librarianship in the next Century. *Library trends* 34 (4), pp. 777-787.
3. Evelyn, H. Deniet (1986). Educating the Academic librarian for a new role as information resource manager. *The journal of Academic Librarianship* 11 (6), p. 364.
4. Raghavan, K.S. etal (1990). Curriculum for Master's programme in Information Science, *Seminar on curriculum development for Information Technology,* Madras, Feb. 20-22.

5. Raju, AAN and Laxman Rao, N. Ed. (1986). Seminar papers on universe of knowledge, its structure and development, its relevance in the curriculum of M.L.I.Sc. Course, Hyderabad : Osmania University, Dept. of Library and information science.

6. U.G.C. (1993). Report of the curriculum Development centre in Library and Information Science.

7. Robert, V. Williams (1990). Specialization in the education of information professionals in *Encyclopaedia of Library and information Science* 45 (10), p. 339.

8. See, J. (1994). Technology ant outcome-based education: connection in concept and practice. The computer teacher 21, pp. 30-31.

9. Thomas, Renee and Muptman, R. (1982).Education for librarianship in the developing countries : A Radical Departure. *Libri* 32 (2), p. 94.

CONTINUING EDUCATION PROGRAMMES (CEPs) FOR AGRICULTURAL LIBRARIANS

INTRODUCTION

The Library and Information professionals of agricultural information systems who have completed their formal library education years ago and who have not subsequently participated in informative educational activities are a dead wood to the agricultural librarianship unless they update themselves. It is essential to organize refresher/in-service training courses for them both at local and national levels.

CONCEPT OF CONTINUING EDUCATION PROGRAMME

Continuing education is based on the philosophy of life long education. It is applicable to all persons irrespective of their age, sex, location and occupation. Paul Lengrand in his introduction to life-long education has mentioned some factors which have created the need for continuing education. They are: Acceleration of change, demographic expansions, Evolution of scientific knowledge and technology and threat of technology unemployment. The above factors are applicable to the agricultural librarianship. In this age of knowledge and information explosion one has to acquaint himself with the latest information and techniques by CEP in one or the other way. The agricultural librarians like the agricultural scientists are supposed to continue the education in their respective fields in order to keep themselves aware of the latest information and knowledge to increase their professional competence.

CONTINUING EDUCATION PROGRAMMES AND AGRICULTURAL INFORMATION PROFESSIONALS

Continuing education is essential for agricultural librarianship and can take many forms – which may or may not lead to degrees. Attending professional meetings and annual conferences of professional organizations, Attending short term workshops, Visiting other libraries and education institutions are some of the CEPs for agricultural libraries. Such CEPs help them to keep abreast of all information in era of knowledge explosion. The following CEPs for agricultural information system may be offered.

FORMAL EDUCATION

(a) **Formal Courses:** Library schools will take responsibilities in offering formal courses. They aim as:

1. Guiding for research.
2. To enable the librarians to hold administrative posts.

(b) **Extension Courses:** Summer schools run by Indian Universities offer extension courses. They provide practical demonstration of different works in library.

(c) **Refresher Courses:** Library schools and national library associations which provide refresher courses are meant:

1. To brush up all and recall acquired Knowledge.
2. Increasing of work efficiency.

(d) **In Service Training**: Agricultural information systems organize in-service training programmes to improve knowledge and to increase job efficiency.

Informal Education Obtained Through:

(a) Reading, writing and editing.

(b) Attending lectures, Professional meetings, Conferences and workshops.

NEED FOR CEPs FOR AGRICULTURAL LIBRARIANS

The new developments are of a high technical nature and librarians trained in the conventional library schools are not trained in the know-how of the latest technology and its application to libraries. Hence in addition to formal library education, there is an urgent need to provide in-service training for library and information science professionals on the new concepts, skills and technologies.

OBJECTIVES OF CONTINUING EDUCATION PROGRAMME

The objectives of the CEP may be the following:

1. To help the library professionals to keep abreast of the new developments in the field.
2. To help them to be fit and competent in the ever changing technology field.
3. To help them to improve their career prospects.
4. To help them in dissemination of new ideas and concepts.
5. To help them to render user-oriented services.
6. To update professional knowledge.

CONSTRAINTS

The success of CEPs for agricultural librarians depends on availability of highly trained man-power and adequate financial resources.

(a) Manpower

Teachers from Library and Information science would have to be drawn to implement the programme. Since they may have expertise in specific areas. This may include reprography, computer software, A/V materials, indexing methods etc. If necessary the teachers from other disciplines are to be selected for conducting CEP for eg. Operation Research Models, Statistical methods, System Analysis, Information Technology etc.

(b) Financial Resources

Adequate financial support from national and international organizations become necessary to conduct CEPs effectively.

UGC, ICAR, DST should take up the responsibilities in providing funds for CEPs.

For offering continuing education programmes at international level, the international organizations such as UNESCO, FAO, IFLA, FID, etc, should take up the responsibilities in providing funds. The state government and Central Government should draw the plan for implementation of CEPs for agricultural librarians. Apart from that the agricultural universities should develop their own financial resources to finance the CEPs for agricultural college libraries.

CURRICULUM FOR CEPs FOR AGRICULTURAL LIBRARIANS

A specific curricula should be framed for the CEP for the agricultural librarians.

1. Fundamentals of Agricultural Science Subjects.
2. Agricultural Library Terminology.
3. Agricultural Library Management.
4. Agricultural Information Processing and Retrieval.
5. Introduction to Agricultural Information Science Resource Sharing and Networking.
6. Micro Computer, Programming, Database Management and Agricultural literature search & retrieval (CD-ROM search)
7. Introduction to Agricultural Information Sources, reprographic services and Inter-Library Exchange Programmes.
8. Agricultural Information Research Methodology and Biostatistics.
9. Practical on:
   - (a) Micro computer Operation-Database Management
   - (b) CABI, AGRIS
   - (c) E-Mail, Internet
   - (d) Desk Top publishing

CONCLUSION

It may be said that library profession is fast changing. Continuing education of library professionals has become a necessity in view of new developments. The Professionals themselves must be alert and keep with the changing time otherwise they become irrelevant and incapable of providing services to the seekers of information. The parent bodies, employers of library professionals must adopt a liberal attitude towards the professionals so that they may update their professional knowledge and skills through continuing education.

REFERENCES

1. Gupta, B. (1978). Continuing education of library and information specialists. Lib. Sc. Doc. 15.
2. Koenig, M.E.D. (1986). Education for information management competition or cooperation ?. Library Trends (Spring).
3. Mangla, P.B. (1997). Library and information science Education in India for the 21st Century; Trends and Issues. Information and society, ed. L.S. Ramaiah. et al.. New Delhi: Ess Ess publications.
4. Mangla, P.B. (1979). Manpower development for information handling, J.Lib. Info. Sc. 4.
5. Seetharama, S. and Gopinath, M.A. (1972). Continuing education for librarians: Its needs and methods: DRTC annual seminar 10.
6. Srivastava, M.K. (1984). Role of Library Associations in Continuing Education to Libraries Lib er. 23.
7. Steig, Margaret F. (1992). Change and Challenges in Library and Information Science Education. Chieago: American Library Associations.

CONSORTIA FOR AGRICULTURAL INFORMATION IN ANDHRA PRADESH

INTRODUCTION

Library Consortia aims at providing materials, information and services by different types of libraries to the needy users. These libraries *may* be in different jurisdictions but agree to serve one another on the same basis as each serves its own constituents.

The agricultural information as a vital resource for agricultural scientists and farmers need not be emphasized when we consider the increasing number of information generators and users. We can visualize to certain extent the magnitude of the problems of collection, organization and dissemination of agricultural information. Agricultural libraries in Andhra Pradesh are playing a vital role in this direction. However, with the growth of agricultural information, there are very often delays in providing service to the end users. The cost of providing service is also rising, mainly because of adoption of manual methods and duplication of investments in holdings.

The above have made it imperative to share library resources and to go for consortia of agricultural Libraries in Andhra Pradesh.

LIBRARY CONSORTIUM

It is equally true, as well as applicable, for all types of libraries/information centers not to hold the full stock of information resources or to procure all information, which may be in demand by its clientele. To solve this problem library cooperation started long ago, such as, library networks, ILL, document delivery, library consortium etc. which are internationally accepted, but at present the more accepted system of resource sharing is library consortia that has come into existence with a wide coverage.

Library consortium is a cooperation between two or more information agencies which have formally agreed to coordinate, cooperate or consolidate certain functions to achieve mutual objectives. It is an association of a group of libraries to achieve mutual joint benefits. Consortia may be formed on a local, regional, national, or international basis; on a functional or format basis on a subject basis.

THE SALIENT FEATURES FOR LIBRARY CONSORTIUM ARE

1. To eliminate the different problems faced by the libraries in order to provide various services to the users.
2. To meet the thrust of information of the vast people due to rapid growth of population all over the world.
3. To cope up with the newly generated knowledge published in different forms, such as, printed and non-printed documents, electronic media on various disciplines, multi-disciplinary and newly generated subject areas; and
4. To collect all the documents published at national and international level, because of the library financial crunch.

FACTORS NECESSITATING CONSORTIA IN AGRICULTURAL INFORMATION SYSTEM IN ANDHRA PRADESH

The following are the factors which influence the agricultural information system towards consortia:

1. Increase of books and journals prices.
2. Limited library budget.
3. Increased cost in providing services.
4. Exponential growth of information in Agriculture.
5. Inadequacy of library materials required for single agricultural library.

6. Increase in variety and degree of users needs.
7. Development of technology.

SHORTCOMINGS OF THE PRESENT SYSTEM

In the present situation, the services of agricultural information system have been found to be far from satisfactory. The drawbacks observed are mainly on account of the following.

1. Non-availability of materials.
2. Lock of sufficient funds.
3. Managerial problems.

NEED FOR AGRICULTURAL INFORMATION CONSORTIA FOR ANDHRA PRADESH

The extraordinary growth and dynamics of information gives rise to financial pressure upon agricultural libraries and information centers to coalesce into consortia in order to share the resources rather than duplicating them. Soaring of cost of reading materials as against tighter budget is also one of the reasons for consortia of library and information system.

The agricultural scientists of various disciplines who are engaged in the field of research, teaching and extension are finding it difficult to keep themselves abreast of new developments within their areas of specialization. It is recognized that current agricultural information relevant to agricultural scientists is dispersed in a large number of research reports, periodicals, and other documents. Added to these the range of variation in information needs, level of sophistication, and capacity for effective application of information among the potential users (famers) will result creating crucial and difficult task for the agricultural scientists to have access to timely, reliable, and precise information.

The present environment in the State of Andhra Pradesh in relation to the availability of agricultural information sources to the various agricultural scientists will make us to think on two aspects of planning on agricultural information systems. Primarily, on the accessibility and awareness of agricultural information sources available in various agricultural libraries and research stations of Andhra Pradesh State. Secondly, by providing nascent information generated and available outside the State, i.e. providing access to national and international agricultural data bases required by the agricultural scientists of the Andhra Pradesh State.

ESTABLISHMENT OF MANAGEMENT COMMITTEE

A consortium is to be established in Acharya N.G. Ranga Agricultural University, Hyderabad. It acts as a Coordinating Centre for participating libraries. It can be formed with the representatives of Agricultural College libraries, Agricultural Research Institution/Station Libraries. The Management Committee may consists of 15 members with the Vice-Chancellor of Acharya N.G. Ranga Agricultural University as the chairperson and at least two Associate Deans (on rotational basis for two years), five members from the Agricultural Research Institutes (on rotational basis for two years), 4 Senior Librarians from different Colleges on rotational basis (again two years), and a representative from Indian Council of Agricultural Research. The University Librarian of Acharya N.G. Ranga Agricultural University will act as Member-Secretary for the Consortium.

The consortia covers, ANGRAU Central Library, two ANGRAU Regional Libraries, seven ANGRAU College Libraries, six Regional Agricultural Research Stations, forty nine Agricultural research Stations, eight Agricultural Research Institutes established by Indian Council of Agricultural Research (ICAR) and Government of India on specific disciplines of agriculture and one International Agricultural Research Institute namely ICRISAT.

FUNCTIONS OF CONSORTIUM COORDINATING CENTRE

The Consortium Coordinating Centre, would function under the management committee for inter-institutional coordination. The important activities of the consortium coordinating centre would be as follows:

- To act as *nodal* agency for increasing the cooperation amongst the participating institutions.
- To *coordinate* all activities concerned with subscription of e-resources on behalf of the consortium.
- To liaison with electronic publishers to provide training and technical help to the participating member institutions.
- To coordinate with Indian Council of Agricultural Research and participating institutions for subscription to resources.
- To organize the meetings of the Management Committee and to decide upon the policy issues.
- To maintain a Web site for the Consortium for the benefit of its members and to encourage sharing of resources in an online mode.
- To propagate the consortium with other institutions and enroll new members in consortium.
- To organize annual meetings of the consortium members.

FUNCTIONS OF CONSORTIUM

Every agricultural library differs from one another according to its collection, information needs of users, working methods, source of finance, processing of information, etc. The various steps of jobs/functions can be adopted by a consortium for effective functioning by standing on a common platform, which are:

AGREEMENT FOR ESTABLISHMENT

A concrete *agreement needs* to be established for the participating libraries in consortia to achieve a common target. All libraries should work in such a manner that they are mutually exclusive, but they have to follow the common mission of a consortium, as a whole. Hence the Management Committee should be formed.

ADMINISTRATION OF LIBRARY CONSORTIUM

To run the consortium smoothly, a statutory body is very much essential to be formed taking library Heads from every library, the university Librarian, Acharya N.G. Ranga Agricultural University as Chief coordinator. Each library will inform about their collection, databases, user services, and training of human resources for handling new equipment, etc. to the Chief Coordinator

FINANCIAL CONTROL

Due to financial crisis, libraries form a consortium for providing better service to the users. The crucial activities of the members of management committee are to take action on the following financial points like:

- Whether a consortium funds be created to subscribe for the core journals in different subjects in multiple copies at a discount rate.
- How to manage the fund of a consortium and how much money is to be collected from each library as a contribution.
- Acharya N.G. Ranga Agricultural University shall take the responsibility to negotiate with some publishers for procuring the resources required by the consortium members.
- Which Standard accounting methods have to be followed for the fund raised by the Consortium.

JOINT WORK

As per guideline of a consortium, individual library will prepare a list of title for resource sharing among participating libraries looking into the user demand and may also initiate to contract with different publishers for negotiation. All the participating libraries under library consortia must work jointly just like a joint venture in business sector to make the consortia a total success. Each library will subscribe the core journals and prepare the union list of titles to be covered under resource sharing programme and the member libraries may also send full content page service of all issues of the title to another library.

EVALUATION

After a certain period, the whole activities of a consortium (i.e. individual as well as co-operative) must be evaluated on the basis of the pre-determined objective. If there is any lacuna in between the work performed and to be performed should be adjusted by taking corrective measures in activities in future. If there is a need for any change in the path of action, it should be changed immediately to achieve the objectives of a consortium.

REFERENCES

1. Biswas, Bidhan Ch. and Dasgupta, Swapna K. (2003). "Opportunities for libraries in managing and resource sharing through consortia: A new challenges for Indian librarians". In Mapping Technology on libraries and People-CALIBER -2003, pp. 291-97.
2. Kaul, H.K. (1999). Library resource sharing and networks. Vigro publications, New Delhi.
3. Murthy, TAV. (2002). Sharing and consortia for India. Processing, IIT Seminar, Kharagpur, pp. 14 – 15.
4. Veeranjaneyulu, K. and Rajagopal, T. (2001). Networking of agricultural information centers in Andhra Pradesh (AP-AGRINET) : A model internal library Movement 23(3), pp. 162 - 176

INFORMATION SEEKING BEHAVIOUR OF AQUA FARMERS OF ANDHRA PRADESH: A CASE STUDY

INTRODUCTION

Aqua farmers are the ultimate users of the aquaculture research findings through proper extension media. The aqua-farming category includes illiterate as well as literate people. The extension workers and rural libraries have a significant role to play in communicating new ideas and practices to the aqua farmers. Aquaculture scientists should consider it as one of their primary duties to impart their knowledge to the farmers. This would be possible only if they convey their findings in popular languages. Aquaculture in Andhra Pradesh has been progressing fast since 1975. At present Indian Crap culture is done in 1.2 lakh hectors, Tiger prawn culture in 80,000 hectors and freshwater prawn culture in 40,000 hectors with 175 hatcheries, 65 processing plants providing employment to 1.6 million people. Aquaculture has significantly increased socio-economic development of the rural people in nine costal districts of Andhra Pradesh. So aquaculture has played a significant role in poverty alleviation, protein food supplement, and employment generation, besides earning foreign currency through exports. At present aquaculture annual export earnings are Rs.3,500 crores from Andhra Pradesh alone.

OBJECTIVE

This study was conducted in order to find out:

(a) various types of information sources of aqua farmer's interest with regard to seed, feed, fertilizers, chemicals and marketing;

(b) to know the various ways through which the aqua farmers seek their information;

(c) to analyze the role of mass media like Radio and T.V. in educating the farmers with the latest developments;

(d) to know aqua farmer's preference regarding newspapers and farm magazines published in Telugu.

METHODOLOGIES

The data required for the study was collected from primary as well as secondary sources. The primary data was collected through questionnaires by interview method. The study was conducted in Nellore District of Andhra Pradesh during January 2004 to June 2004. 150 literate aqua farmers were selected for the purpose of this study. As the Nellore District is having three Revenue Divisions, 50 samples were selected from each Revenue Division.

ANALYSIS OF DATA

The analysis of the data is presented below:

The table 1 reveals the various sources used by the aqua farmers for getting information regarding seeds, fertilizers, pesticides and marketing. Majority of the respondents i.e. 71 per cent get the required information through friends, 59 per cent of respondent get from relatives and 57% from the representatives of Private companies.

Table 1: Source of Information used by Aqua Farmers

Sl.No.	Source	Total Respondents	Responses	% of response
1.	Friends	150	107	71
2.	Relatives	150	89	59
3.	Newspapers	150	72	48
4.	Radio	150	78	52
5.	Television	150	63	42
6.	College of Fishery Science	150	48	32
7.	Wall Posters	150	56	37
8.	Representatives of private companies	150	86	57
9.	Farmers meetings	150	82	55
10.	Fisheries Department	150	79	53

Thus the main sources of information for getting aquaculture information are friends, relatives and representatives of private companies.

It is evident from the table 2, that majority (55%) of farmers get the required information from farmers meetings.

Table 2 : Telugu Newspapers as Sources of Information

Sl No.	Source	Total Respondents	Responses	% of response total respondents
1.	Eenadu	150	82	55
2.	Vartha	150	66	44
3.	Andhra Jyothi	150	26	17

Information by reading *Eenadu* and 44 per cent get from *Vartha* newspaper.

Table 3 : Telugu Farm Magazines as Sources of Information

Sl No.	Source	Total Respondents	Responses	% of response total respondents
1.	Aqua International	150	26	17
2.	Annadata	150	42	28
3.	Aqua Tech	150	9	6
4.	Padipantalu	150	4	3
5.	Simhapuri Rythu	150	65	43

The table 3 shows that 43 per cent of aqua farmers get aquaculture information by reading *simhapuri Rythu* and 28 per cent from *Annadata*

Table 4 : Aqua Farmers and Air Braodcasts

Reply	Respondents	% of Respondents
Yes	56	37
No	94	63
Total	150	100

The table 4 shows that only 37% of Aqua Farmers hear AIR broadcasts and remaining respondents do not do so.

Table 5 : Type of Popular Pragrammes

Sl No.	Pragramme	Total Respondents	Responses	% of response total respondents
1.	Polampanulu	56	14	25
2.	Padipantalu	56	6	11
3.	Palleseemalu	56	52	92

The table 5 shows that most of the aqua farmers (92%) hear *Palleseemalu* programme.

Table 6 : Aqua Farmer's Programmes on T.V.

Sl No.	Name of the T.V. Channel	Programme	Total Respondents	Responses	% of response total respondents
1.	Doordarshan	Gramadarshini	150	108	72
2.	ETV	Annadata	150	64	43
3.	ETV2	Jai Kisan	150	72	48
4.	Teja	Rythu Mitra	150	124	83

It is evident from the table 6 that majority of aqua farmers watch Teja T.V. programme 'Rythu Mitra' and Doordashan programme 'Gramadarshini' for their imfromation needs.

Majority (83%) of the farmers are satisfied with the usefulness of the T.V. pragrammes. Only 17 per cent respondents have negative opinion, may be due to lack of non-availability of T.V (Table 7).

Table 7 : Aqua Farmers' Opinion regarding Aquaculture Programmes on T.V.

Reply	Respondents	% of Respondents
Useful	124	83
Not useful	26	17
Total	150	100

FINDINGS

The main findings of the study are:

1. The analysis shows that the aqua farmers are getting aquaculture information regarding seeds, fertilizers, pesticides and marketing mainly from friends, relatives and representatives of private companies.
2. It is clear from the analysis, that the respondents are reading only two newspapers (Eenadu and Vartha) for aquaculture information.
3. The respondents are getting aquaculture information by reading mainly two farm magazines (Simhapuri Rythu and Annadat).
4. The analysis shows that the "palleseemalu" programme is very popular among the farmers because of it being broadcast in the evening time.
5. 83% of the farmers watch "Teja T.V.- Rythu Mitra" programme because it is very popular among the watching in the evening time.
6. Majority of the farmers are satisfied with usefulness of the T.V. pragrammes.

SUGGESTIONS

Different measures to be taken for improving access to needed information by the aqua farmers are:

1. All India Radio and T.V. Channels should broadcast aquaculture related programmes only in the evening time and not in the daytime, as the farmers are busy with their work during the day;
2. The Agricultural University should establish Agricultural Information Communication centers for fulfilling the information requirements of aqua farmers.
3. Under the prevailing circumstances the most efficient way of providing access to information to farmers is by establishing rural information centers as integral part of public library system;
4. The Information and Public Relations Department and Fisheries Department should open information centers in every Mandal and regularly screen programmes on the latest developments in agraculture;
5. The scientist and extension workers should visit villages regularly to know problems and tune their research accordingly;
6. Kisan Melas should be conducted in each Mandal with big publicity once in a year; and
7. The Government should start Krishi Vighan Kendras and Farmer' Training Centers in every district.

APPLICATION OF INFORMATION TECHNOLOGY IN ACHARYA N.G.RANGA AGRICULTURAL UNIVERSITY LIBRARY SYSTEM

INTRODUCTION

Computer applications in the field of library and information science has been one of the major factors attracting the attention of the information professionals throughout the world. The emergence of microcomputers and the software available for same have made computer application economically viable.

The library world is being acclimatized to computer environment in both library routines as well as information storage and retrieval. Computerization has retrieved the pressure of work load in libraries to a great extent. It offers efficient services and opens a new era in bibliographic control. It also provides access to required databases in the country and abroad. The computer is capable of storing information for longer period of time in minimum space in the electronic-readable form. This information is capable of being recalled with ease and utmost speed, as and when warranted. Achieving progress of research in agricultural sector and proper improvement in production is mainly dependent on the supply of information to the agricultural scientists. The automation of agricultural libraries is very essential to satisfy the information needs of agricultural scientists.

APPLICATION OF COMPUTERS TO AGRICULTURAL LIBRARIES

The facilities intended to be provided by application of computers to agricultural libraries are as follows :

1. Library house- keeping operations
    - Acquisition
    - Cataloguing
    - Serials Control
    - Circulation Control

2. Information Retrieval
    - In-house Catalogue
    - Online access
    - CD-ROM Discs
3. Library Network and Networking
4. Office Automation
    - Word processing Applications
    - LAN (Local Area Network)
    - WAN (Wide Area Network)

AGRICULTURAL RESEARCH INFORMATION SYSTEM (ARIS)

It is not difficult to envisage all the components of information technology which are used in an agricultural research and extension institutions in developing country like India. While there are undoubtedly barriers for developing countries, not the least of which will be the costs, the potential for using advanced information technology is undeniable, and undoubtedly opens up new vistas for the transfer of scientific information. It is the ICAR who instituted computerization of its Research Institutes, State Agricultural Universities as well as head quarters with the aim of improving research management by bringing information technology culture to scientific community. Keeping this in mind a project called Agricultural Research Information System (ARIS) is being implemented with the financial aid of World Bank under NARP.

Besides creating the infrastructural facilities ARIS have four information modules namely.

1. Agricultural Management Information System (ARMIS)
2. Agricultural Personal Information System (ARPIS)
3. Agricultural Financial Information System (ARFIS)
4. Agricultural Library Information System (ARLIS)

ANGR Agricultural University obtained required information Technology infrastructure facilities under ARIS project and AHRD Project.

ANGR AGRICULTURAL UNIVERSITY LIBRARY SYSTEM

With a view to cater the information needs of teachers and students as well as research and extension personnel in the three geographical regions of the state, the library system was reorganized in 1980 locating Central Library at Hyderabad and Regional Libraries at Tirupati and Bapatla and College Libraries at other campuses. The library system consists of the following :

Central Library : Central Library, Hyderabad

Brach Libraries : College of Veterinary Science, Hyderabad

Extension Education Institute, Hyderabad

Regional Library : ANGRAU Regional Library, S.V. Agricultural College, Tirupati.

ANGRAU Regional Library, Agricultural College, Bapatla

College Libraries : College of Home Science, Hyderabad.

College of Home Science, Bapatla.

College of Agricultural Engineering, Bapatla.

Agricultural College, Naira.

Agricultural College, Mahanandi.

Agricultural College, Aswaraopet.

College of Fishery Science, Muthukar.

College of Veterinary Science, Gannavaram.

The libraries functions under the overall control of the University Librarian who is one of the statutory officers of the University directly responsible to the Vice-Chancellor. The Library Committee consists of the Deans of faculties and the Associate Deans of the Colleges. The

Vice-Chancellor being the Chairman and the University Librarian as the Member Secretary. The Central Library is managed by the University Librarian and the regional Libraries and college libraries by the Assistant Librarians.

INFORMATION TECHNOLOGY INFRASTRUCTURE

The information Technology infrastructure in ANGRAU Libraries are as follows :

CENTRAL LIBRARY

LAN Server

Internet Server

CD-Net Server

Computer -486,Pentium, Pentium II and Pentium III.

Laptop

Laser Printer

Dot Matrices Printers

Inkjet Printer

CD-Writer

HP-Scanner

Bar Coding Machine

Fax

Nodes-18

Modem

REGIONAL LIBRARIES

LAN Server

CD-Net Server

Computer -486, Pentium and Pentium III

CD-Writer

Laser Printer

Dot Matrices Printers

Inkjet Printer

Modem

COLLEGE LIBRARIES

Pentium III

Dot Matrices Printers

Inkjet Printer

Modem

SOFTWARE

LIBSYS

WINDOWS 1995, 1998 and 2000

Novel Netware

MS Office 2000

INFORMATION TECHNOLOGY APPLICATION IN ANGRAU LIBRARIES

The study is intended to reveal the present use of information technology in ANGR Agricultural University Libraries. In order to put the results of this study into a framework that can be more easily assimilated, a wide range of possible IT application have been grouped into following pattern:

1. Library house-keeping operations
    - Acquisition
    - Cataloguing
    - Serials control

2. Information Services
   - OPAC
   - Internet Services
   - E-Mail Services
3. Library Networking
   - Local Area Networking
   - Wider Area Networking
4. Office Automation
   - Word Processing
   - Accounting

INFORMATION TECHNOLOGY FACILITIES IN ANGARAU LIBRARIES

SI.No	Name of the Library	Computer Facility	Internet Facility	CD-ROM Services	Xerox Facility	LAN Facility
1	Central Library Hyderabad	Yes	Yes	Yes	Yes	Yes
2.	Regional Library, Tirupati	Yes	Yes	Yes	Yes	Yes
3.	Regional Library, Bapatla	Yes	Yes	Yes	Yes	Yes
4.	College of Home Science, Hyderabad	Yes	Yes	No	Yes	No

SI.No	Name of the Library	Computer Facility	Internet Facility	CD-ROM Services	Xerox Facility	LAN Facility
5.	College of Home Science, Bapatla	Yes	No	No	Yes	No
6.	College of Agricultural Engineering, Bapatla	Yes	Yes	No	Yes	No
7.	Agricultural College, Mahanundhi	Yes	Yes	No	Yes	No
8.	Agricultural College, Naira	Yes	No	No	Yes	No
9.	Agricultural College, Aswaraopet	Yes	No	No	Yes	No
10.	College of Fishery Science, Muthukur	Yes	Yes	No	Yes	No
11.	College of Veterinary Science, Gannavaram	Yes	Yes	No	Yes	No

COMPUTERIZATION

The Central Library and Regional Library have built up a database of dissertations submitted to ANGRAU with a view to providing pin pointed and exhaustive bibliographic information. The Central Library and

Regional Libraries provides CD-ROM based database searches by subscribing the CABI. The computerizations of day to day activities are in progress.

TRAINING STAFF

All the library staff have been trained under AHRD Project on Application of Computers to Library and information Science in India and Abroad.

PROBLEMS IN APPLICATION OF I.T. IN ANGRAU LIBRARIES

The main problems experienced by the ANGRAU Libraries in its computerization are as follows :

1. The retrospective conversion of bibliographical information into machine readable form.
2. Lack of enough trained personnel to maintain the database.
3. Lack of staff orientation towards computerization.
4. Lack of motivation.
5. Lack of knowledge and training to the users.
6. Lack of awareness about the possible benefits of IT application.
7. Lack of understanding between library staff.

A PROFESSIONAL INSIGHT INTO COMBATIVE CONSORTIA ON SUBSCRIPTION TO E-JOURNAL TOWARDS COST EFFECTIVE SOLUTIONS FOR UNIVERSITY LIBRARIES

INTRODUCTION

India has a large educational system comprising thousands of Colleges and hundreds of Universities along with several institutes of higher learning and centers of excellence. The university libraries, being the nerve centers of the higher education and learning, play a vital role in support of all the activities such as teaching, research and publication. Our higher education system is passing through lot of problems due to economic recession and cost recovery concepts. The university libraries in India are faced with the challenges to meet the user requirements due to various reasons such as exponential growth of knowledge, price escalation, and budget restrictions. The increasing growth in the enrollment of students and researchers, lack of proper and adequate infrastructure further aggravates the overall problems and challenges for the college and university libraries. This situation has paved the way for serous thinking on capabilities to compensate for reduced budgets. E-Subscription is one of the emerging toolkits for libraries of survive in the present circumstances.

SUBSCRIPTION ADVANTAGES

1. Speed of access to the latest Information.
2. Ability to implement multimedia elements.
3. Quick searching.
4. Linking from and to other resources.
5. Security.
6. Single user or location restrictions did not apply.
7. Down loading an article to a personal computer for later use or printing is easier than photocopying or scanning.

PROSPECTS OF CONSORTIA BASED SUBSCRIPTION E-JOURNALS

(a) The pooling of expertise in resolving the complicated tasks. For example, implementation of modern network applications.

(b) The saving of precious time of the personnel by centralization of some of the common activities.

(c) The realization of financial savings through joint acquisitions and the dissemination of document databases as well as in administrative and management work and

(d) A stronger position vis-à-vis vendors, whether in price negotiations or in other business dealings.

Consortia based subscription to electronic resources is a part of 'Library Consortium' and it can be a viable solution in order to increase the access to electronic resources across institutions at a lower cost.

POINTS FAVOURING OPTIMAL COST OF CONSORTIA SUBSCRIPTION

1. Optimal pricing.
2. Reduced overheads.
3. Simpler budgeting.
4. Easier apportioning or allocation of costs.
5. Facilities on early transition to becoming a profit center.
6. Fixed price tenures.
7. Deeply discounted print options.
8. Archival and perpetual access possibilities.
9. Differential pricing models.

INTEROPERABILITY- THE KEY ISSUE

Interoperability is a concept that addresses the extent to which different types of computers, networks, operating systems and application work together effecting exchange information in a useful and meaningful manner. The technology-enabled library Consortia environment is heterogeneous; it hosts to many different technologies various data, multiple applications and other networked resources. A functional goal is to hide this heterogeneity from users so that they may effectively search for information, communicate and perform other tasks.

ADDITIONAL MOTIVATING FACTORS FOR COMBATIVE CONSORTIA

Though there are several factors a few of them are enumerate below.

1. Periodical discussions among libraries, which are involved in Library Consortia.
2. Development of Inter Library lending code for regional, state, local and other special groups.
3. Adequate photocopying and microfilming facilities in all libraries.
4. Efficient postal and communication facilities.
5. Availability of union catalogue of holdings of all libraries.
6. Development of subject socialization.
7. Standard cataloguing format should be adopted; and
8. Promptness in supply of materials.

COUCLUSION

University Libraries in India are facing challenges due to budget crunch, reduced staff, devaluation of rupee, steep hike in conversion of foreign currencies, and escalation in cost of publication.

The UGC-Info net is also exploring the possibilities to subscribe e-journals through a consortium to benefit college and University Libraries in the country. Though consortia based subscription to electronic resources can help college and University Libraries in order to enhance their access capacity to large number of periodicals yet there is a need to develop a policy for consortia based subscription to electronic resources which could clearly indicate the terms and conditions favoring and protecting academic interests of University Libraries.

REFERNCES

1. Arora, Jagadish (2002). Electronic publishing: An overview. In training programme on digital library content creation, access and management. IIT Delhi 18to 22 December 2001. Delhi IIT
2. Becker, Joseph and Becker, J.F. (1983). Technology trends in Library Resources sharing networks Australian Special Libraries; News16(4), pp. 121-124.
3. Bhatt, R.K. and Madhusudhan, M (2004). University Libraries in India and Electronic Journal : The role of Consortia based subscription of E-Journals for the effective use of financial resources. Second International Caliber2004, New Delhi, pp. 129-133.
4. Liew, C. L. et al.(2000). A study of graduate end users, use and perception of e-journals Online information review 24 (4) pp. 302-313.
5. Chawla, Ashok and Viswanath, M (2004). Vendors role in the sustenance of the Consortia subscription to E Resources: A case study of IEEE/IEE- Electronic Library-IEL in the INDEST Consortium. ICDL-2004.
6. Viswanath, M (2004), Sustenance of the consortia subscription to E Resources: Role of matrics. Second International Caliber 2004, New Delhi, pp. 136-140.

CHALLENGES FOR AGRICULTURAL LIBRARIANS IN 21st CENTURY

INTRODUCTION

The mission of the libraries i.e. providing excellent information services to patrons has never changed. Libraries are now more known as information centres while adapting the changing technology. Technology has changed the way the librarians serve their users and it will continue in future also. While continuing to provide many traditional information services, librarians are developing new skills and taking on new roles that are necessary to support hybrid information services.

Information technology has its impact in all areas of library work such as providing user access to digitized resources in the local environment and also remote access. It is in this context that libraries must develop individual solutions that are appropriate to local circumstances. Many libraries have added electronic materials and services to the traditional items associated with physical space. These resources available via the web include e-books, online catalogues, licensed databases, e- journals, freely available Web resources and local digital collections.

The need to provide value-added library services in Agricultural Libraries is all the more necessary to help in academic, extension and research pursuits. The user expectations have increased with the proliferation of electronic sources. Libraries have to redesign their services on the context rather than form and delivered to the users. The information is available electronically through networks and consortia 24 hours a day. In fact Technological developments and the impact of Information Technology in agricultural libraries have led to the emergence of a new concept called 're-engineering' .

The 21st century is going to witness a tremendous growth in Virtual Libraries or libraries without walls or global Libraries.

EVOLUTION OF LIBRARIES

The Table 1 shows the metamorphosis of the libraries from the early age to the digital era.

Table 1

Evolution of Libraries

Paper Library **Collections=Paper** **Bib Control= Paper**	**Automated Library** **Collections=Paper** **Bib Control=Electronic**	**Electronic Library** **Collections=Digital** **Bib Control=Digital**
1965-1995		1995 - date

Traditional Libraries Vs 21st Century Libraries

The differences between traditional libraries and 21st century libraries are mentioned in Table 2.

Table 2

Traditional Libraries	**21st Century Libraries**
Documents	Internet
User goes to Library	Library goes to the users
OPAC	Web based OPAC
Documents (Books, Journals etc.)	E-Resources (E-Books, E-Journals etc.)
Cataloguing, Indexing & Classification	Search Engines & Metadata
Circulation (Issues & Returns)	Access direct with Internet facility
Resource Sharing	Consortium

RECENT TRENDS IN AGRICULTURAL LIBRARY SYSTEM

The past two decades have seen a great deal of change due to Information Communication Technologies (ICT) resulting in a demand for new patterns of scholarly information. These technology advancements have made significant impact on the growth of knowledge and in unlocking the human potential.

The following are the recent technological trends in Agricultural Information System

1. Automation
2. Digitalization
3. Internet
4. Consortia
5. RFID Technology
6. Web Designing for Libraries
7. E-Resources
8. ICT in Libraries
9. Creation of Institutional Repository
10. Learning Resources

CONSORTIUM FOR E-RESOURCES IN AGRICULTURE (CeRA)

India is predominantly an agrarian country, and the growth of agriculture is reflected in the good yields of different crops that depend on various factors – natural and man-made. Agricultural research, the backbone of agricultural growth in the country, demands timely dissemination of knowledge being generated and is being updated across the globe from time to time. Hitherto, R and D institutions have been procuring print versions of journals and literature in aid of science and technology. With the advent of internet facilities and advancement of web technology, almost all reputed international journals are available on-line and can easily be accessed by researchers over the network. Since ICAR is having network connectivity across the institutes and state agricultural universities, select journals could be made available over the network for the use of scientific community. Keeping this broad objective in mind, the National Agricultural Innovation Project (NAIP) has funded for establishing the Consortium for e-Resources in Agriculture (CeRA) at the Indian Agricultural Research Institute (IARI).

Objectives of CeRA

- To develop the existing R & D information resource base of ICAR institutes/ Universities etc., comparable to that existing in world's leading institutions/ organizations.
- To subscribe e-journals and create an e-access culture among scientists/teachers in ICAR institutes/agricultural universities.
- To develop a Science Citation Index (SCI) Facility at IARI for evaluation of Scientific publications.
- To assess the impact of CeRA on the level of research publications measured through SCI.

Work Plan

The work plan to meet the above objectives comprises

- The establishment of a CeRA Co-ordination cell for negotiation and subscription of e-journals.
- Formation of mandatory Committees for guiding and monitoring the work of the Consortium.
- Installation and maintenance of dedicated server and e-storage facility for maintaining archival repository of e-resources in electronic form.
- Development/adoption of a suitable software for e-access.
- Identification of e-journals publishers/vendors and procurement of e-journals.
- Organization of workshops/ trainings to facilitate and popularize e-access, and
- monitoring and impact analysis of e-resources through web of science and NASS ID.

The Consortium consists of the following organizations under National Agricultural Research System (NARS) :

Sl.No.	Organisation	No.
1.	Deemed Universities	5
2.	ICAR Headquarters and Krishi Anusandhan Bhawans	3
3.	National Bureau	5
4.	Institutes	42
5.	National Research Centres	21
6.	Project Directorates	9
7.	State Agricultural Universities	38
	Total	**123**

- Document delivery systems are to be developed at IARI, NDRI, IVRI and CIFE to enable those Consortium partners who do not have the print versions of particular scientific articles.
- Periodic reporting of the activities of the Consortium from time to time.

Expected Outcome/Impact/Deliverables

- Online accessibility of all important journals related to agriculture and biotechnology to researchers and students of the Consortium members.
- Quick access to world R & D information as available world wide.
- Permanent archive of the subscribed e-databases.
- Improvement in the quality of scientific publications, and teaching and research guidance.
- Providing one print version of each title subscribed for ready reference at IARI free of cost.

CeRA plays an important role in agricultural libraries and success of consortia totally depends upon the members of consortia. The agricultural libraries should also make use of the available technology to meet the expectations of user information requirements. An agricultural library consortium plays a vital role in India providing information to agricultural scientists in right time. It helps to increase production and to achieve self-sufficiency.

HUMAN RESOURCE DEVELOPMENT IN AGRICULTURAL LIBRARIES IN 21ST CENTURY

The application of modem technology allows us to access and store information quickly. The availability of computers has transformed the agricultural libraries. The information technology provides a wide range of services of which agricultural libraries could avail themselves to supplement existing ones.

It is necessary to impact latest information technology skills and techniques to agricultural librarians to acquire such latest practical skills along with their professional knowledge. Otherwise it would be difficult for them to cope up with the situation in the fast changing global scenario. At times they may have to lose their ground for survival into cognizance. Agricultural library professionals should equip themselves with the new IT skills and practical knowledge in delivering the information services to users.

Suggestions to Human Resource Development in Agricultural Libraries

1. Training programmes should be conducted for agricultural library staff to equip themselves with the application of latest information technology.
2. ICAR and UGC should provide necessary funds to acquire necessary information technology equipment.
3. Library & Information Science Schools should organize in-service training programmes regularly.
4. The agricultural library professionals should be encouraged to use modern technologies.
5. User Orientation Programmes should be conducted for readers to enable them to use the library resources efficiently and effectively.

6. Specialized and short term courses have to be conducted by the INFLIBNET and ICAR to agricultural library professionals to update their knowledge.

NEW ROLE OF AGRICULTURAL LIBARIANS IN 21ST CENTURY

The Agricultural Librarians should provide the following new services in the 21st century.

1. Offering virtual reference services.
2. Choosing resources & managing licensing.
3. Creating metadata elements.
4. Collecting and digitizing archival material.
5. Maintaining digital repositories.
6. Information literacy.

CHALLENGES FOR AGRICULTURAL LIBRARIANS IN 21ST CENTURY

The present day environment leads to sophisticated and challenging tasks to Agricultural Libraries. The following challenges are facing by the agricultural librarians in 21st century.

1. Changing formats environment of information material
2. Technological Challenges
3. Changing Role of Librarians
4. Challenges in collection of Virtual Information
5. Archiving or preserving the Virtual Information
6. Challenges in Accessing Virtual Information
7. Special skills needed for Librarians to face the challenges in building a virtual library.

The Agricultural Librarians should anticipate and integrate new technologies to redesign access and delivery mechanisms to maximize utilization of information and to minimize the costs of supporting such services. They should improve their skills in the use of Information Technology to overcome above challenges to provide efficient and effective information service to the agricultural scientist.

CONCLUSION

Application of ICT have made a great impact on agricultural libraries and information centres, and the computers and associated facilities have come to be recognized as indispensable tools for accurate storage, processing, retrieval and dissemination of information. All agricultural librarians should be familiar with the benefits of modern technology and perform efficiently. The challenges are to design library services and to manage operations exploiting the advantages offered by ICT. To that end it is pertinent to understand the process of technology development and technology diffusion in general and ICT in particular.

Today Agricultural Libraries are using web technology extensively for user services to provide dynamic services to the remote users. There is a need for Agricultural Librarians to get extensive exposure with the Information Communication Technology in the 21st century to provide effective information services to the agriculture scientist to increase food production and to achieve self- sufficiency.

REFERENCES

1. Biswas, Bidhan Ch and Dasgupta, Swapan K (2003). Opportunities for libraries in managing and resource sharing through consortia: A new challenges for Indian libraries". In Mapping technology on Libraries and people -CALIBER -2003. pp. 91-97.
2. Kaul, H.K. (1999). Library Resource Sharing and Networks. Virgo publications, New Delhi.
3. Murthy, T A V. (2002). Sharing and consortia for India. lIT Seminar, Kharagpur. pp. 14-15.

4. Veeranjaneyulu, K and Rajagopal, T. (2001). Networking of agricultural information centres in Andhra Pradesh (AP-AGRINET) : A model. International Library Movement 23(3). pp. 162-176.

5. Vithal, K.S.R and Veeranjaneyulu, K. (2008). In Management of E-Resources in Academic Libraries edited by Kosteswara Rao, M. and Veeranajneyulu, K., B.S. Publications, Hyderabad. pp. 101-110.

GIGANTIC ROLE OF E-BOOKS: AN OVERVIEW OF FUTURE LIBRARIES

INTRODUCTION

Electronic books are one way to enhance the digital library with global 24 × 7 accesses to authoritative information, and they enable user's access specific research material easily, quickly and effectively. Library users are also able to remotely search, locate and check out e-books from the libraries using online public access catalogs (OPACs). As with any new opportunity new challenges emerge and utilizing the internet to deliver book content is no exception. Integrating e-books into the digital library has created challenges and opportunities for LIS professionals; publishers and e-books providers.

E-BOOKS

An e-book is based both on emulating the basic characteristics of traditional books in an electronic format, as well as leveraging internet technology to make access to e-books easy and efficient to use.

Electronic books offer creative possibilities for extending access as well as changing learning behavior and academic research. Content can always be accessible regardless of time or place, to be read on Pc or on portable book readers, books need never go out of print and new editions can be easily created. One can carry several times at once on a portable reader and over time build a personal library, features such as full text searching, changeable font size, mark up, citation creation and note taking will enhance usability. Print text can be integrated with multi dimensional object, sound to create a whole new kind of monographic work.

Despite the considered pronounce of e-books, our text force contacted that all the element that world make the e-book market viable are not quite in place. Element that we consider important to store regarding academic use of e-books.

1. Content

2. Software and Hardware standards and protocols.
3. Digital rights management
4. Access
5. Archiving
6. Privacy
7. The market and pricing
8. Enhancement and Ideal e-book futures.

Content

Academic readers need to rely on authority and integrity of content. E-books content should match any print version and include all its elements text groups and illustration. We anticipate that academics will increasingly incorporate information, that can only be presented as non-print media, content needs to be separated from the transferable in a non proprietary format, into a variety of software and hardware readers, both to offer readers a choice of additional features, and to make it possible for libraries to loan e-book content.

Software and Hardware standards and protocols

(i) Software

Non commercial monographic electronic text for scholarly exchange have been available on the internet since that early year of UNIX file exchange, file transfer, gopher and finally type text transfer protocols.

Currently there is no established standard for an interoperable e-book format for commercially produced e-books that address publishers needs to support commercial end user distribution and that also enables world value for the consumer.

(ii) Hardware

A pragmatic factor in using e-books is the reading and using them, yet e-books hardware devices are still not quite practical or cost effective enough to progress very deeply into the market. A variety of devices are being developed to replicate some of the virtues of printed monographs, including portability and network independence, so that e-books themselves will function on a variety of platforms. The two basic kinds of e-book readers on the market to-day are the full sized and the palm sized reader. Currently all readers use proprietary file formats.

Digital right management

One of the most critical element in the development of electronic publishing that will impact libraries involves digital right management still being developed. DRMS are either hardware or software that enforce control over intellectual property, such as limit by user time, fee, and for extent of content.

Access

In addition to standards allowing interoperability and ownership through digital rights management issues of access include user awareness and the ability to accommodate simultaneous users. So that researches can readily identify e-books, libraries must be able to integrate with other format in catalogs and integrated library systems, in standardized forms of bibliographic information and in metadata such as MARC and other appropriate metadata.

Archiving and long term Access

Two of the roles collection of academic libraries are building research level collectors and acting as archiving of research information. Currently e-books vendor purchase models allow some flexibility such as a premium price for perpetual access versus more modest pricing for annual access to a revolving group of lines.

Privacy

Some e-book vendors are creating individual user account to track how many times an individual has changed on, and to create my library type features. The individual account on the vendor's web site may infringe on privacy, since it would, be possible for vendors to report exactly what an individual had accessed regarding the vendor's system.

The market and pricing

The demand for simultaneous access to individual e-book times should not regime libraries to purchase multiple copies of e-book and pricing models should be developed that permit same level of access.

Enhancement and Ideal future

This functionality may be varied inclusion of multimedia information. Full text searching mark-up citation formatting, reference linking conventions portability, interoperability on a variety of devices, availability in advance of print, advantages one pricing and the ability to share or hour information, given the variety of user needs, non-proprietary interoperability of e-book content will be needed.

ADVANTAGES OF E-BOOKS

(a) The biggest advantage is the so called every hoariness of an e-book. In a similar fashion, a student away from any research library can access that library's honoring via a computer attached to a medium.

(b) A large state or region can medium their enhance on multiple copies of a book and instead make available the text as an e-book.

(c) The e-book can be made available as a CD-ROM and occupy a fraction of the space of a paper volume of the same title.

(d) The e-book either in the online or CD-ROM form may contain animation and live-action illustrations not possible in printed books.

E-BOOK OPPORTUNITY

The internet had sensed an emotion in the book publishing industry with the emergence of the e-book.

- Easy access to content.

- On demand availability.
- Prevention from being lost, stolen, or damaged.
- Capability to search within a book and across a coalition of books.
- Acridity to be linked to other resources including deaconries and the saurian
- Absence of physical space management
- Device independence for accessing the content.
- Access to content using standard web browsers
- Customizable search interfaces.
- Easy transportation and
- Access from any where and any time

E-BOOKS AND FUTURE OF LIBRARIES

In spite of many challenges, progress has been made in the production and distribution of e-books during the last two years. Libraries, publishers, e-book providers and integrated library system vendors have worked together to implement and integrate equations systems, text version collection development strategies propose and adopt new revised and combined standards provide new e-book hardware and software identify and test new indexing methods for full text e-books test new access and usage models; and interact archiving policies and procedures for e-books.

CONCLUSION

The role of e-book in academic libraries is still not clear, and there is considerable development of standards technologies and pricing model needed to make the market for e-books viable and sustainable. Technologies for reading and using e-books are not yet convenient enough for the longer text format to have made much market panoramas.

Print has many rights and powers that e-books don't. We like e-books but we must not allow ourselves to be look into technology or legal/social paradigms that impair our patch discourse of our community.

REFERENCES

1. Carty, Susan (2000). The digital text book is coming. Information Searcher 1/01, pp. 123-127.
2. Crawford, Walter (2000). Wine models, one name: Untangling the e-book, muddle. American libraries, pp. 56-59.
3. Dillon, Danni's (2000). Digital books; making them, work for publishers and Libraries. College and research libraries 61, pp. 391-393.
4. Dillon, Dennis (2000). "E-books; the UT-Aushri Experience. Texas Library Journal 76, pp. 112-115.
5. Gibbons, Susan 2001. E-books; some concerns and surprises; Portal; Libraries; and the Academy 1, pp. 71-75.
6. Hawkins, Donald T (2000). E-books; A Major publishing revolution part I; General considerations and issues; online 24, pp. 14-28.
7. Halter, Donnie Small (2000). e-Books in Libraries : Some early experience and reaction, Searcher 8, pp. 63-5.
8. Lynch, Clifford (1999). Electrifying the book. Library Journal 124, pp. 3-6.

ROLE OF INFORMATION TECHOLOGY IN TRANSFER OF AGRICULTURAL TECHNOLOGY

Information systems are assuming an ever-increasing importance in the agriculture development and playing an important role for sweeping changes in socioeconomic development of the country. Computer based information systems/databases and communication networks are today a pre-requisite for taking coherent and balanced decisions. Computer is a major component of information technology (IT) which is helpful in doing routine things faster, more accurate and leaves us to concentrate on the analysis of the information and proper decisions. The use of computers has brought awareness on the potential of IT in agriculture. Area of usage advanced from computing to the existing multitude of uses in agriculture research. Decision Support System (DSSs) Expert Systems (ESs), Extension Services to farmers, Farm management, training etc.

To sustain self-sufficiency in food it is essential that the advancement in IT must be extended to farming community. Information services to the farmers can be improved through the creative use of IT. Farmers use information on weather, markets, package and practices of new varieties in various crops. IT with its various tools such as Internet GIS, ESs and DSSs can play a vital role in improving the existing status of the farmers by providing various information needed to them.

INFORMATION TECHNOLOGY AS A MEDIUM FOR AGRICULTURAL INFORMATION TRANSFER

The extension of the basic computer technology includes new and powerful peripheral devices and the development of the computer as a resource for the storage, management and retrieval of information in all of its manifestations. IT industry can offer solutions, which can improve the working situations and lives of the farmers within this country. Establishing computer-aided and internet connected Rural knowledge centers will helps to trigger a knowledge revolution in agriculture and will lead to an efficient and eco-sensitive precision farming movement.

The trends in information Technology have brought about a new set of services and have transformed the traditional services in terms of fast delivery. Government around the world has taken advances to provide better services to farmers, citizens and businesses. District, towns and large villages have become nodes to offer information services, banking services credit and insurance services. It is important to note that they have become hubs for agriculture, commerce, revenue collection, transportation etc. All of these have created Net based co-operative.

In 70 villages in Kolhapur and Sangli districts of Maharashtra Internet booths have been established. The wire village project has been jointly implemented by Government of India through National Informatics Centre (NIC) and warana co-operative Society. In Kerala all the villages are getting linked to citizens to compare the development priorities of their villages in the state. In Andhra Pradesh and in Dhar district of Madhya Pradesh villages have begun to file applications and follow their progress on net (Including land transfer) world Tel. Reliance and the Tamil Nadu government are jointly laying 3000 km. of optical fibers to create a Tamil network, which will offer online services in different fields.

LEVERAGING INFORMATION TECHNOLOGY FOR AGRICULTURE

The information technology can surely provide some long-term solutions to the challenges facing by the agricultural sector. Over the last few decades the impact of IT has undoubtedly impacted for the growth of agriculture sector. Information communication technologies such as the internet, Mobile Phones have dynamically changed the way we live and our social consciousness. Internet provides us a best tool for the information sharing and mutual communication. The potential of contribution of information science to effective and stable production by several models such as crop growth prediction and decision support is also very high.

Recent developments in information Technology have created opportunities for access to information in quantity and quality, which has never been available to scientists earlier. Equipped with this resource it is now possible to design and implement innovative and powerful technological packages for gaining uncommon benefits. It is also not necessary to start from the beginning to reach the stage of reaping benefits; technologies and the related material can be acquired through quick technology transfers by using modem channels of communication.

INTERNET IN TECHNOLOGY TRANSFER

The latest advances in information technology (IT), computers, telecommunications and internet have provided conductive environment for adopting new technologies and making the method of instructions more effective and interactive. In spite of facilities available cost effectiveness of the advanced media and its suitability, the potentiality of the electronic media is not being used for transferring the recent technologies properly in agricultural sector. According to the reports of the extension cell of ICAR, New Delhi, 80 per cent of the technology generated by the researchers in agricultural sector are either not transferred to the ultimate users or they do not find applicability in the farmers field. Out of the remaining 20 per cent of the technologies, around 50 per cent do not impress the framers due to mainly its effective technology transfer that ranges from language to effective demonstration using the available tools. Now internet has emerged as the most powerful IT tools for storage and retrieval of information. It works around the clock and connects every nook and corner of the globe. With an unprecedented growth in the quantum of knowledge worldwide and easy accessibility, internet has become an unavoidable necessity for organizations especially those who involve in technology transfer. Internet and its application offer enormous potential in enhancing the efficiency of implementation. Monitoring and evaluation of various development programmes of the agricultural research Education, Extension, Planning of crop production forecasting marketing etc. it can also act as a powerful medium for the transfer of modern technology to the farmers.

The consequences of these major changes in the information generation storage and dissemination are that there is a greater need than ever before in the history to develop coordinated approaches in the agricultural sector to identify, organize and make available information in time to the end-users when it is required in user friendly format.

ROLE OF INTERNET IN TECHNOLGY TRANSFER

We all aware that internet can play an important role in the efficient use of available resources to maximize agricultural production and productivity. Internet offers enormous potential in enhancing the efficiency of implementation, monitoring and evaluation of the various development programs of the Government. It can also act as a powerful medium for the transfer of modern technology to the farmers. Internet and its application are also very useful in agricultural research education, extension, planning of crop production, forecasting post harvest management, marketing disaster management etc. Such application of internet would ensure a proper growth of the agricultural sector on a sustainable basis. Besides, farmers can also benefit directly through the utilization of various web sites that have been launched by the Government. Generally, the collection of information on various aspects of crops/agriculture is done by the local patwari or his representative who is already preoccupied with many activities most generally they fill the records of their own without even a visit to the site. The collected information does not provide the real picture of yield and other related issues. For this purpose trained personnel be engaged who are capable to use internet especially database designed for the purpose of storing the data collected at mandal level; these data can be pooled for all the mandals on district level that will give comprehensive information for each block and district. Through internet this information can be retrieved for the entire state and country. Internet web sites are another tool for the farmer to acquire information and solution. Both the development of agriculture and co-operation and ICAR has programs of the Government web sites can be visited on http://agricrop.nic.in and http://www.icar.org.in, respectively.

These are the following attributes of internet information

(a) A global encyclopedia virtually on any topic – from very technical to very triviat

(b) Accessible 24 hr per day, or whenever the farmer has time or the in and no need for marking a meeting appointment.

(c) Information Overload – more information than anyone can possible the time or the energy to sort through.

(d) Quality of information – tremendous variations depending upon the (perspective) of the source organization for posting the information (environment lobbyist legislator research, social commentary, product sales)

(e) Technical level of information – ranges from general public information press release extension information research report production information.

Authenticity/accuracy of information – there is no governing internet regulates the integrity of the displayed information.

Agricultural information can be made round the clock through the internet is emerging as a potential tool to contribute to agricultural development sites and rural portals in regional languages can be developed and hosted world wide web through internet. This will provide information to farmers local language besides Chat software's in local languages can be development and made available to the farmers through these rural portals the software's farmers in getting into discussions with experts and other farmers and experience persons and will help in exchanging views and information and find solution problems. This will facilitate in providing quick and correct information Multimedia can be used to present the information in variety of ways e.g. information in the form of graphics, animation, audio and video clipping can be available through interact available through internet. Since farmer's needs vary with season, crop e- commerce site based on these parameters can be developed to agribusiness services needs of farmers.

REFERENCES

1. Devraj, Chaturvedi and Kare, A.P. (2001). Information technology agricultural extension. Agricultural Extension Review 13(3).
2. Farooqi, Samir; Singh, Niranjan and Islam, Shahnawazul (2002) Information Technology : Potential and prospects. Agricultural Extension Review
3. Sandhu, A.S. (1993). Textbook on Agricultural communication and methods. New Delhi : Oxford & IBH Publishing.
4. Swaminathan, M.S. (2001). Science and our agriculture future Agricultural Situation in India.

KNOWLEDGE SHARING IN AGRICULTURAL RESEARCH: INFORMATION DISSEMINATION INITIATIVES OF I.C.A.R.

INTRODUCTION

The application of Information & Communication Technologies in agricultural practices and research has given a new lease of life to all the stake holders of agricultural knowledge producers and users, viz. agricultural scientists, entrepreneurs, farmers, policy makers and of course librarians and information professionals. There is a sense of interconnectivity among all these players nationally and internationally, giving rise to agricultural knowledge revolution that increased the extent, intensity and speed of agricultural knowledge production and use. As we know "knowledge is power" and "knowledge management" is a powerful tool which helps to realize, share, nurture this power for the benefit of our scientific and farming community.

Libraries of ICAR Institutions and SAUs are playing a vital role in providing the required information to their users for education, research and extension. The capacity building of these libraries is the need of the hour for efficient and effective delivery of library & information services at NARS.

The aim of the National Agricultural Innovative Project (NAIP) is to facilitate accelerated and sustainable transformation of Indian agriculture in support of poverty alleviation and income generation by collaborative development and application of agricultural innovation by the public research organizations in partnership with the farmer's group, the private sector, the civil society organizations and other stakeholders. The NAIP project has sanctioned the following projects for knowledge sharing among Agricultural Libraries in India:

1. Consortium for e-Resources in Agriculture (CeRA)
2. Krishi Prabha : Indian Agricultural Doctoral Dissertations Repository.

3. Strengthening of Digital Library and Information Management under NARS (e-Granth).
4. E-Publishing and knowledge system in Agricultural Research (E-PKSAR)

CONSORTIUM FOR E-RESOURCES IN AGRICULTURE (CeRA)

India is predominantly an agrarian country, and the growth of agriculture is reflected in the good yield of different crops that depend on various factors, both natural and man-made. Agricultural research, which is the backbone of agricultural growth in the country, demands timely dissemination of knowledge which is being generated and updated across the globe from time to time. Hitherto, R & D institutions have been procuring print versions of journals and other documents in aid of science and technology. With the advent of internet facilities and advancement of web technology, almost all reputed international journals are accessible on-line by researchers over the network. Since ICAR has the network connectivity across all its institutes and state agricultural universities, select journals could be made available over the network for the use of scientific community. Keeping this broad objective in mind, the National Agricultural Innovation Project (NAIP) has funded for the establishment of the Consortium for e-Resources in Agriculture (CeRA) at the Indian Agricultural Research Institute (IARI).

Objectives of CeRA

- To develop the existing R & D information resource base of ICAR institutes/ Universities etc., comparable to that existing in world's leading institutions/ organizations.
- To subscribe e-journals and create an e-access culture among scientists/teachers in ICAR institutes/agricultural universities.
- To develop a Science Citation Index (SCI) Facility at IARI for evaluation of scientific publications.
- To assess the impact of CeRA on the level of research publications measured through SCI.

Work Plan

The work plan to meet the above objectives comprises:

- The establishment of a CeRA Co-ordination cell for negotiation and subscription of e-journals.
- Formation of mandatory Committees for guiding and monitoring the work of the Consortium.
- Installation and maintenance of dedicated server and e-storage facility for maintaining archival repository of e-resources in electronic form
- Development/adoption of suitable software for e-access
- Identification of e-journals publishers/vendors and procurement of e-journals
- Organization of workshops/trainings to facilitate and popularize e-access and
- Monitoring and impact analysis of e-resources through web of science and NAAS ID.

The Consortium consists of the following organizations under the National Agricultural Research System (NARS) :

Sl.No.	Organizations	No.
1.	Deemed Universities	5
2.	ICAR Headquarters and Krishi Anusandhan Bhawans	3
3.	National Bureau	5
4.	Institutes	42
5.	National Research Centres	21
6.	Project Directorates	9
7.	State Agricultural Universities	38
	Total	**123**

- Document delivery systems are to be developed at IARI, NDRI, IVRI and CIFE to enable those Consortium partners who do not have the print versions of particular scientific articles.
- Periodic reporting of the activities of the Consortium from time to time.

At present CeRA is providing full text articles of the following publishers:

- Springer Verlag :
- Annual Reviews
- CSIRO Publishers
- Open J- Gates
- Elsevier Publishers
- Taylor and Francis
- Indian Journals.com

Most of the agriculture related journals from selected publishers are available in CeRA. At present, more than 2900 journals are accessible by the researchers.

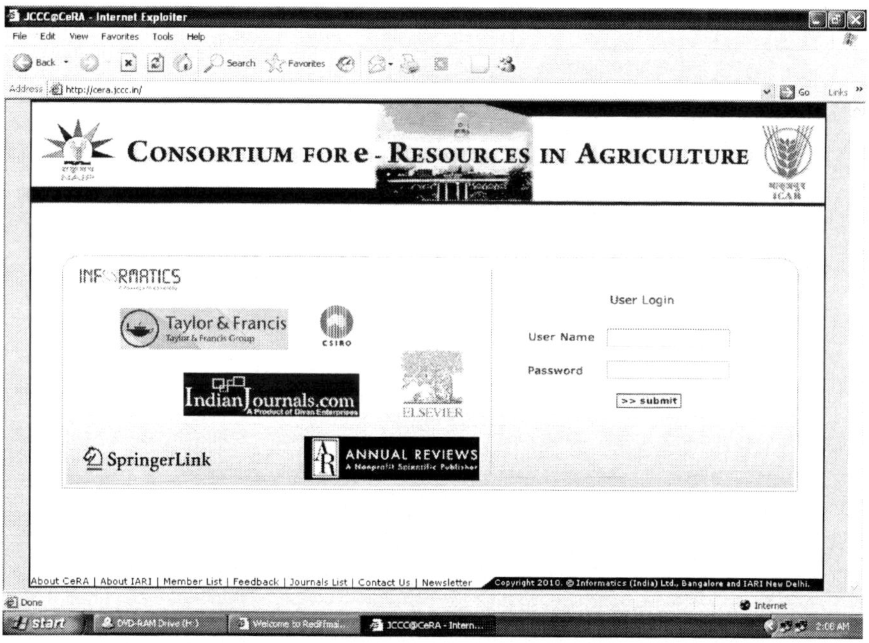

Expected Outcome/Impact/Deliverables

- Online accessibility of all important journals related to agriculture and biotechnology to researchers and students of the Consortium members.
- Quick access to world R & D information.
- Permanent archive of the subscribed e-databases.
- Improvement in the quality of scientific publications, and teaching and research guidance.
- Providing one print version of each title subscribed for ready reference at IARI free of cost.

CeRA plays an important role in agricultural libraries and success of consortia totally depends upon the members of consortia. The agricultural libraries should also make use of the available technologies to meet the user information requirements. An agricultural library consortium plays a vital role in India providing information to agricultural scientists at the right time.

KRISHI PRABHA: INDIAN AGRICULTURAL DOCTORAL DISSERTATIONS REPOSITORY

Krishi Prabha is a full text database of Indian Agricultural Doctoral Dissertations submitted by research scholars to the 45 State Agricultural Universities and Deemed Agricultural Universities between 01-01-2000 and 31-12-2006. This database, lists about 10,500 Dissertations, created by Nehru Library, Ch. Charan Singh Haryana Agricultural University, Hissar with financial support from Indian Council of Agricultural Research, New Delhi under its National Agricultural Innovation Project.

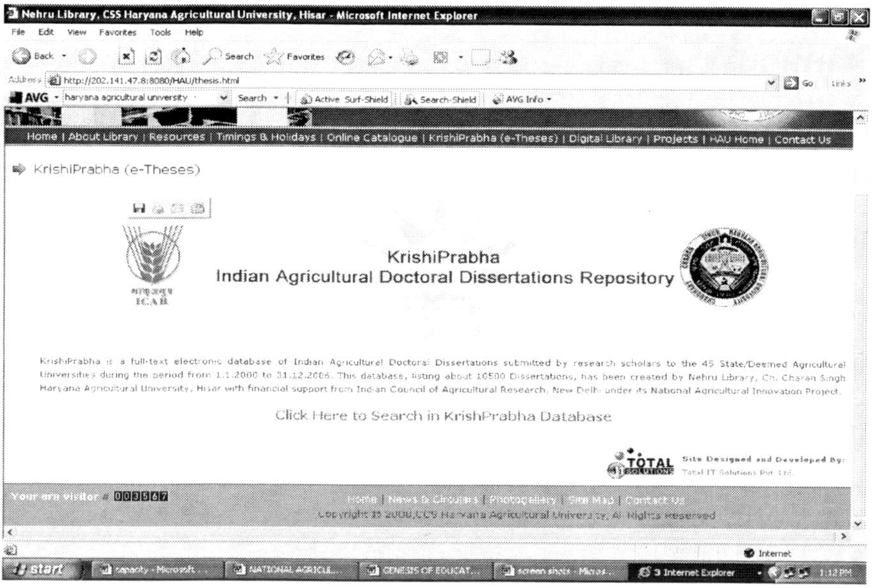

Expected Output/Impact/ Deliverables:

- Integrated Database of abstracts of PhD theses.
- Digitized database of over 10,000 doctoral theses in Agriculture.
- Trained manpower in preparation and management of e-theses.
- Standardized protocol and formats for submission of e-theses by research scholars.

STRENGTHENING OF DIGITAL LIBRARY AND INFORMATION MANAGEMENT UNDER NARS (E-GRANTH)

'Strengthening of Digital Library and Information Management under NARS (e-GRANTH)' is a subproject under component-1 of National Agricultural Innovation Project (NAIP), Indian Council of Agricultural Research (ICAR). National Agricultural Research System (NARS) in India comprises of Agricultural Research Institutions and Agricultural Universities.

It provides digital access to library resources of 12 different research institutes and agricultural universities which include OPAC, important institutional repositories, rare books and old journals and makes them publically accessible over internet under NARS with Online Computer Library Center (OCLC) partnership.

OBJECTIVES

1. To create Online Public Access Catalog (OPAC) under "Indian Agricultural Research Group Catalogue" of all 12 library resources with Online Computer Library Center (OCLC) partnership.
2. To digitize important institutional repositories (limited to IARI, New Delhi; IVRI, Izatnagar; ANGRAU, Hyderabad and UAS, Bangalore) including rare books and old journals and make them open access under NARS.
3. To strengthen capacity building for library and information management system (open to all libraries of NARS).

Union Catalogue and Digital Libraries are the new paradigms of modern library management. The National Agenda for Governance, which is the Government's policy blueprint, has taken due note of the ICT Revolution that is sweeping the globe. Project Gutenberg, Google Book Search, Windows Live Search Books, Internet Archive, the Library of Congress World Digital Library, the digital library at the University of Michigan, and the Million Book Project completed in partnership with India. China and USA are considered leaders in the field of digital archive creation and management. Many Indian Libraries have taken up the challenge of modernization to develop the **MA**chine **R**eadable **C**atalogue (MARC) and machine readable full text documents to provide greater accessibility to full text databases of different nature, changing the focus of public libraries from providing library services to the fulfilling the long pending focus of

Right To Information. The National Knowledge Commission has recently recommended an Open Access mandate for all publicly funded research.

Indian **N**ational **A**gricultural **R**esearch **S**ystem (NARS) has a very large collection of institutional repositories in agriculture and allied sciences, spread over the country in different libraries of ICAR Institutes and **S**tate **A**gricultural **U**niversities (SAUs). Digitization of these valuable archives would allow online access to researchers, teachers and students to which they would not otherwise have access. Duplication of record creation can be avoided by pooling the efforts through common electronic protocols. Union Catalogues are useful to librarians, as they assist in locating and requesting materials from other libraries through interlibrary loan service. Subscription to **Online Computer Library Center (OCLC), USA** would enable our libraries to be discoverable by the network of global library system and sharing of online resources more effectively.

1. Indian Agricultural Research Institute (IARI), New Delhi
2. Indian Veterinary Research Institute (IVRI), UP
3. University of Agricultural Sciences (UAS) , Bangalore
4. GB Pant University of Agriculture & Technology (GBPUAT), Uttarakhand
5. Ch. Charan Singh Haryana Agriculture University, Hisar
6. Acharya NG Ranga Agricultural University (ANGRAU), Hyderabad
7. National Dairy Research Institute (NDRI), Karnal
8. Central Institute of Fisheries Education (CIFE), Mumbai
9. CSK Himachal Pradesh Krishi Vishwavidyalaya (CSKHPKV), HP
10. Mahatma Phule Krishi Vidyapeeth (MPKV), Maharashtra
11. Tamilnadu Veterinary and Animal Sciences University (TanuVAS), Tamilnadu
12. Directorate of Information & Publication of Agriculture (DIPA)

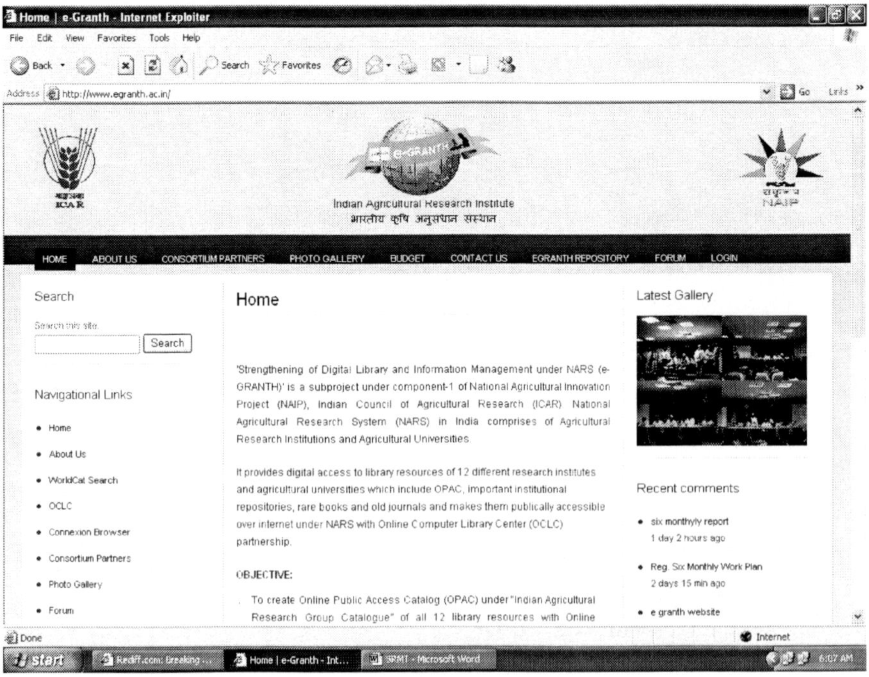

Expected output

- ➢ Partner libraries would become at par with world libraries with respect to library and information management rich with digital resources and web content.
- ➢ It is hoped that annual conference of libraries and information centres of NARS in the lines of 'CALIBER' would become a regular event.
- ➢ The libraries and librarians would be ready to migrate to modern management paradigms with web 2.0 and library 2.0 technologies.

E-PUBLISHING AND KNOWLEDGE SYSTEM IN AGRICULTURAL RESEARCH (E-PKSAR)

The project 'E-publishing and Knowledge System in Agricultural Research' is expected to provide crucial information for accelerated and sustainable transformation of Indian agriculture through print and electronic mode. The objective is to provide improved communication link among information generators, i.e. public research organizations, and its users, viz. researchers, farmers' groups, Panchayati Raj Institutions, private sectors and other stakeholders, for poverty alleviation and income generation.

The access to scientific and technical information is essential for economic and social development. The research results are mainly communicated and validated through publications in scientific journals. ICAR periodicals play an important role in increasing public awareness about the agriculture scenario by communicating national agricultural information on regional problems with international implications in English and national language. They also help in development of local capacity in editorship, publishing and writing. However, there is need to increase their credibility to promote research on local problems and increase usage of science in decision making. Also, there is an urgent need for a platform where ICAR periodicals can be hosted, to give them better international visibility. The ICAR periodicals fulfill all the requirements of international standards, but mainly face a drawback of limited visibility and the solution is e-publishing, coupled with online access.

Electronic publishing holds immense potential for the agricultural research information dissemination. The advantages include improved global as well as national reach, speed of publication, available search services, rapid access and additional capabilities. This project will strengthen the ability of the ICAR to act as a catalyzing agent for rapid transformation of Indian agricultural research system by strengthening the research information dissemination process.

CONCLUSIONS

Use of the latest information & communication technologies in information handling is the need of the hour for all libraries. Knowledge sharing can only be achieved through global networking and the use of the Internet for creation and dissemination of research information effectively to all the stake holders.

CeRA, e-Granth, Krishi Prabha and E-Publishing are some of the initiatives taken up by the ICAR to create a knowledge sharing network for effective dissemination of Indian agricultural information which is otherwise distributed all over the country. The agricultural libraries are expected to make use of the available technologies to meet their user information requirements. A national consortium of agricultural libraries like CeRA can play a vital role in providing research information to scientists located at any place. All these initiatives are expected to help increase the country's agricultural production and thereby achieve self-sufficiency in food.

REFERENCES

1. http://cera.jccc.in
2. http://hau.ernet.in
3. http://icar.org.in
4. http://.naip.icar.org.in
5. www.egranth.ac.in
6. Veeranjaneyulu, K and Rajagopal, T (2001). Networking of agricultural information centres in Andhra Pradesh (AP-AGRINET) : A model. *International Library Movement* 23 (3). Pp. 162-176.

MANAGEMENT OF E-RESOURCES IN INDIAN LIBRARIES

The ever increasing complex needs of the information seekers on one side and the information explosion on the other side, followed by the ever shrinking library budgets, price escalation etc., have become the serious problems in providing adequate and suitable documents/information services by the library and information centres. This had led the collection building with application of Information and Communication Technology (ICT), which helps in accessing the electronic resources, and sharing the same in our country. Consortium is mainly formed for sharing of e-resources.

The rapid growth of e-resources and the complexity in managing these resources has posed new challenges for the LIS professionals and hence requires to be addressed properly. Now a days some vendors are working on providing complete solution for E-resource management but these are also not comprehensive solution to these problems. However some alternative steps can also be taken by the libraries to manage their e-resources properly.

NEED FOR E-RESOURCES IN LIBRARIES

There are several reasons to justify the collection building of the electronic resources in libraries. They are as follows.

- Ever increasing cost of the print journals.
- The long time gap between publication of a journal and its availability in the market.
- Technological development.
- Easy to access and search.
- Low searching time.
- Availability of vast amount of the e-resources in every field of knowledge.

SELECTION OF E-RESOURCES

The common types of e-resources that will be considered are

- Databases
- Electronic books/monographs
- Electronic full-text journals
- Internet Sites
- Reference Tools
- Teaching/Education materials

The following primary and secondary criteria should be evaluating all types of e-resources.

Primary Criteria

1. Relevance, Content and Scope
2. Organization of Material
3. Quality of Resources
   (a) Authority
   (b) Content
   (c) Special Attributes

Secondary Criteria

1. Cost
2. Access Capabilities
3. Systems and Technical Support
4. Simultaneous Users
5. Licensing
6. Software requirements
7. Training and Educational Support

E-RESOURCE CYCLE IN LIBRARIES

- Origin, Identification and location
- Testing
- Selection and Acquisition
- Access
- Renewal or Cancellation

E-RESOURCE MANAGEMENT IN INDIAN LIBRARIES

Management of the electronic resources has become comparatively easier with the development and application of Information Technologies in libraries. Nowadays a library can manage its e-resources, effectively by adopting one or more methods. Different methods to manage the e-resources in libraries are discussed below.

Creation of E-Resource Library

With growing emphasis on use of technology, libraries have undergone major structural change in terms of their collection, organization and services. The traditional concepts of libraries are becoming obsolete day by day with the emergence of new 'digital' means of storing and disseminating information. The advanced computing system and the networking revolution have brought a remarkable expansion and abilities to generate and disseminate information in digital formats both on-line and off-line. These technology-based developments have led to speculation on new concepts of libraries i.e. 'Digital Library'. Creation and maintenance of digital library is the best process to manage the digital objects in libraries.

Components of E-Resource Library

The Digital Library mainly comprises of Digital object, Hardware, Software, Live ware as well as Finance.

Digital Objects

Digital library generally consists of digital objects. A digital object refers to digital materials, which are of two kinds - born digital and digitized. Born digital means a material which is produced originally in digital form, some examples are E-books; E-journals, E-resources etc. and Digitized materials are those, which are converted from other formats.

Hardware Requirements

Hardware is the name given to all physical components (largely electronic) from which the computer is made up. Thus we can say that hardware is the collection of equipments.

Software Requirements

Software contains the complete and unambiguous description of each task in terms of the available operations of computers. For running e-resource libraries successfully, there is a need of a well developed software.

Manpower

Although hardware and software are essential components for digital library, efficient and experienced manpower is highly required to run the e-resource library smoothly.

Finance

Financial planning is also an important component in Digital Library. Since establishment of e-resource Library requires a huge amount of money initially, both recurring and non-recurring expenditure should also be taken into account. It means the arrangement of financial support to different heads.

ADVANTAGES OF E-RESOURCES

The advantages of electronic resources in libraries are many. Some are:

- Low cost of production compared to printed documents.
- It allows remote access from any where at any time.

- It can be used by many users simultaneously.
- Provides hyper links to related additional resources.
- It is interactive and allows interaction between author/publisher and users.
- It supports searching capabilities.
- It eliminates printing and postage cost.
- It does not require physical processing.
- It can easily merge with altering service.
- It provides improved access through full text searching.
- It can solve the problems of missing issues of journals.
- It can be updated more easily and reduces storage space.
- It can easily be managed by adding bookmarks and personal notes to the sites or by downloading into private files or databases for coping and editing.

DISADVANTAGES OF E-RESOURCES

Although there are many advantages of e-resources in libraries, there are many disadvantages also. Some are:

- Initial high infrastructure and installation cost.
- Hardware and software compatibility issues between publishers and users.
- Excessive printing of documents.
- Difficulty inherent in relating to a large amount of data on a screen.
- Causes more concern about copyright.
- Efficient manpower is required.

PROBLEMS IN MANAGEMENT OF E-RESOURCES IN INDIAN LIBRARIES

- Selection of E-Resources
- Budget
- Hardware and Software
- Training to Library Staff
- Orientation Programs
- Usage Statistics
- User Feedback

CONCLUSION

Advances in information technology have changed many activities of library particularly the collection development activity. Electronic resources are gradually replacing the print materials of library. In this changing library context, it is highly essential to manage these e-resources properly for their wide and effective use, otherwise a major portion of library budget, which is spent towards this will be a waste. Also information seeking behavior of the users has revolutionized the acquisition, access and use of electronic resources.

E-Resources are gaining lot of momentum in Indian libraries. Indian libraries need to change their collection development policies to acquire e-resources to satisfy the changing information needs of the user community.

REFERENCES

1. Anilkumar, N. (2001). E-journals : Major issues. Information Today and Tomorrow 20 (3). pp. 9-10.
2. Brose, T.R. (2006). Digital Information Resources Management. IASLIC Bulletin 51 (2). pp. 48-53.
3. Tamar, S. Electronic Resource Management Systems : The need and the realization. New Library World 106 (5/6). pp. 208-218.
4. Winter, E and Dykstra, G (2002). E-libraries come into its own. Information Today 19. p.7.

LIBRARY HUMAN RESOURCE DEVELOPMENT IN THE DIGITAL ERA

Once Henry Ford said: *'take out my building, take out my machines and all capital but leave my men with me, I will become Henry Ford again.'*

INTRODUCTION

Libraries are facing tough challenges as creation and communication of the value they create for their stakeholders in today's highly competitive information market requires quite different set of attributes and skills than the current workforce and their organizational culture may possess. Technologically-driven change has had a significant impact on the field of library and information science during the past two decades. Strategies for responding to general shifts in information use have been reflected in all types of library settings. Academic libraries, for example, have developed a variety of new services alongside their users changing ways of interacting with information. The rise of information commons, sometimes also called teaching and learning centers, in many academic libraries is directly related to shifting perspectives on the nature of information service in relation to technology.

HUMAN RESOURCE DEVELOPMENT – CONCEPT

Human Resource Development is a process of helping employees in an organization to acquire new skills and competence on a continuing basis. In developing human resources, employees of an organization are helped to acquire skills, knowledge and other capabilities required by the organization according to its standards in a planned process to meet the present and future expectations of the organizations on human side in a conducive environment with result orientations. This is the responsibility of the organization that he should provide to employees the facilities in a systematic and planned way to :

- Acquire and sharpen skills and capabilities to perform various tasks associated with their present and expected future roles.
- Develop their general capabilities, discover and exploit their inner potential for their own and/ or organizational development purpose.
- Help to bring about an organizational culture in which collaboration among sub-units is strong and contributes to the supervisor – subordinate relationships, team-work and the professional well-being and pride of the employees.
- Facilitate upgrading of skills through mechanism and instruments; and
- Improve abilities by the introduction of strategic and operational HRD planning and its execution.

NEED OF HRD IN LIBRARIES

Human Resource Development or HRD is required for any type of the organization including libraries because of:

1. Survival and Stability;
2. Growth and Development;
3. Change and Diversification;
4. Retuning the activities to become more effective;
5. Providing highest quality in Products and Services;
6. Playing a leadership role and
7. Obtaining goodwill and reputation through user satisfaction.

DIGITAL ERA

ICT is a shorthand for the computers, software, networks, satellite links and related systems that allow people to access, analyze, create, exchange and use data, information, and knowledge in ways that, until recently, were almost unimaginable. The term 'ICT' is used almost interchangeably with the Internet. The Internet together with its applications (the most well known being the World Wide Web) is the infrastructure that brings together people, in different places and time zones, with multimedia tools for information, communication, data and knowledge management in order to enlarge the range of human capabilities.

The world has become a global village. Information management and dissemination are increasingly becoming technologically driven with an exponential rate of turn-over in inventions and creations. This includes all of the resources (men and material) and scenario that characterizes the digital/electronic information management, communication as well as information delivery services especially as found dominant in the 21st century.

ROLE OF LIBRARIAN IN DIGITAL ERA

Librarian is a professionally trained person responsible for the care of a library and its contents, including the selection, processing and organization of information resources and the delivery of information, instruction and loan service to meet the needs of its users. In an online environment, the role of the librarian is to manage and mediate access to information that may exist only in electronic form.

Information resources in all formats with adequate depth and quantity need to be provided in the library for students' experimentation and exploration. Such resources should promote student-centered, resource based, problem-solving and independent learning. Self paced resources that enable the customization of learning and the impacting of specific skills to students should be promoted.

ACTIVITIES OF ACADEMIC LIBRARIAN

New technologies are changing the services that libraries provide; for example, online reference and instruction, desktop document delivery, self-checkout, and user initiated library loan and direct borrowing. It is an established fact that in this part of the world, training and education for librarians are grossly inadequate as evident in the very minimal skilled hands available to man and actualize the transformation from purely traditional to if not purely electronic but at least hybrid library services. An academic librarian holds specialist responsibility for providing information on the subject which is dealt by the organization/institution. The following is a list of tasks typically undertaken by academic librarians:

- managing and developing collections of books, journals and websites;
- managing buildings, furniture and equipment;
- managing staff, which may involve recruitment and selection, appraisals, disciplinary action, staff training, as well as day-to-day management of staff;
- managing budgets and, in some cases, purchasing resources;
- maintaining relationships with external bodies such as publishers, information gatekeepers, suppliers etc.
- contributing to academic course development and liaising with academic departments;
- creating, updating and managing information resources - both electronic and printed; selecting, acquiring and cataloguing information using library and information software;
- assisting researchers with literature searches using databases, printed resources and the internet;
- delivering information and learning skills courses for students and staff;
- dealing with user enquiries which may involve one-on-one advice sessions;

- keeping up to date with relevant debates in the library sector;
- participation in professional groups networks and consortia.

ACDEMIC LIBRARIANSHIP IN PRESENT CONTEXT

New technologies are changing the services that libraries provide; for example, online reference and instruction, desktop document delivery, self-checkout, and user initiated library loan and direct borrowing. It is an established fact that in this part of the world, training and education for librarians are grossly inadequate as evident in the very minimal skilled hands available to man and actualize the transformation from purely traditional to if not purely electronic but at least hybrid library services. The present digital environment leads to sophisticated and challenging tasks to Academic Libraries. The following challenges are facing by the academic librarians in 21st century.

- Changing formats environment of information material.
- Technological Challenges.
- Changing Role of Librarians.
- Challenges in collection of Virtual Information.
- Archiving or preserving the Virtual Information.
- Challenges in Accessing Virtual Information.
- Special skills needed for Librarians to face the challenges in building a virtual library.

The Academic Librarians should anticipate and integrate new technologies to redesign access and delivery mechanisms to maximize utilization of information and to minimize the costs of supporting such services. They should improve their skills in the use of Information Technology to over come above challenges to provide efficient and effective information service.

IMPACT OF ICT IN LIBRARIES

ICT has impact on each and every activity of the library. It has affected staff, materials and buildings. It has impact on books, journals, infrastructure, image, timings, growth, reports, recruitments, promotions, tools and users. ICT has gradually entered into the libraries and made the librarians dependent on it. Its arrival was fast in technical and special libraries than the academic libraries. ICT has brought several changes in the day to day activities of the libraries, even the nomenclature of the libraries have changed to digital libraries, library 2.0, hybrid library etc.

IMPACT ON LIBRARY HUMAN RESOURCE

With the arrival of information technology in libraries the working of library professional changed according to the arrival of various techniques being adopted by them. It created a sort of digital divide in the libraries. The staff working in the libraries got divided into various groups. There is one who are well versed with the latest technology in the libraries and they accepted it happily and they welcomed the arrival of new technology.

Application of ICT has made biggest impact on the library professionals. The profession which was considered as dull, less active, peaceful and relaxing has altogether changed. There was a time when librarian was only concerned with the cataloguing codes, classification schemes, circulation of books, receiving of periodicals, book purchase and enrolment of users. Now the scenario has changed and the should be well versed librarian with web 2.0, digital library, e-books, networks, gateways, RFID, bar-coding, library software packages, OPACs, electronic databases, consortia, digitization, computer hardware etc.

MAJOR IMPACT ON LIBRARIANS

Image and Expectations

Today the librarian is required to know about the best library software, best antivirus, e-database of different subjects. With the arrival of various online sources of information such as e-books, e-journals, OPAC etc. The image of the librarian will only be improved if he provides services without wasting the time of the users. While selecting a person for a job the authorities ask librarian to search databases and provide his academic profile that how much research he has done and what is the level of his publications by using the impact factor. New Academic Performance Indictors (API) provided by UGC for higher scale promotions require high standards research publications. The faculty expects and consults librarian to introduce them with new resources in their particular subjects. Librarians arrange several user awareness programmes for users which has further increased his respect among them.

Librarian and ICT Tools and Techniques

Library professionals have long hours of duty. They are to provide services on holidays and late in the evenings also. After passing out from library schools they get less opportunity and time to involve themselves with further studies. But technological advancement in libraries does not bother about the compulsions and difficulties of the library professionals and it continue to grow and change with time. New tools such as digital technology for digital preservation, digitization, digital cameras, scanners, printers, barcode and RFID techniques, latest library management software, UPS and labs, antivirus systems, knowledge about bandwidth and IP address have forced the librarian to study and know about all these tools and technologies. Now librarians are busy with lot of activities whole day. He is to check regularly about the access of online resources, train new users, watch internet connection and speed, send e-mails to users about latest information products and services, receive requests and look after ICT tools.

MULTI-DIMENSIONAL ACTIVITIES OF LIBRARIAN

Librarian and e-Resources

Now the librarian is like a trainer of fighter pilot who sits along with him and takes him into the unending sky similarly now librarian sits with users and takes them into the unending sources of knowledge from one information resources to another. This can only be done if he himself is well versed with the various e-Resources provided by the different e-Publishers.

Librarian and Authorities

With the arrival of new databases the librarian is required to suggest higher authorities about their need, accessibility, price, authenticity and use. He discusses and negotiates with the publishers on behalf of his parent organization. Today he is in contact with various institutions like UGC, INFLIBNET, NISCAIR, DELNET, and other institutions of national importance for the technological assistance. At times he is to prepare projects to get grants from these institutions.

Impact on Training

Librarian arranges various training programmes of these institutes in his own libraries for the benefit of their staff and other library science professionals to enhance their skills in ever growing ICT era.Library profession deals with the teaching community who's information needs keep on changing from time to time. Librarian has to keep him up to date in order to provide better services.

MODERN LIBRARIANS: SKILLS AND COMPETENCIES

ICT provides an excellent opportunity for library professionals to manage themselves better. All Librarians should be perfectly familiar with the benefit of modern technology and perform efficiently. In order to manage the change library environment, the library professionals needs to be trained in the application of necessary tools in their work environment.

Existing information professionals do not go back to LIS School to learn about all the latest development in the electronic information environment. Short-term courses, Refresher courses, Seminars and

Workshops will help the professionals to learn the new information technologies and latest happenings in LIS profession.

The quality of all the libraries and information services ultimately depends on the caliber, suitability and performance of their staff, properly trained and highly motivated workers are essential.

ICT HELPS LIBRARY PROFESSIONALS

In new technology era, the role of library professional has changed and continues to change from being an instructor to a constructor, facilitator and create learning situation. ICTs can help library professionals in the following ways:

- Encouragement to librarians for developing ICT skills by the respective organization.
- ICTs enable to enhance the initial preparation by giving good teaching.
- With the help of ICT, library professional can access with colleagues, institutions and Universities, centres of expertise, rich resources at cyber space.
- ICTs enable to access online libraries, journals and research to enable individual learning.
- ICTs provide lifelong professional development by providing courses at virtual situation, training on demand, orientation and refresher courses through video conference or online.

LIBRARY PROFESSIONALS AND TECHNICAL SKILLS

The following technical skills are required for library professionals in the digital era.

- Basic knowledge of a personal computer
- Internet Knowledge

- Software Knowledge
- Networking Knowledge
- Hardware Knowledge
- E-Resources Knowledge
- Digitization Knowledge

NEED FOR TRAINING

The application of modern technology allows to access and store information quickly. The availability of computers has gradually changed the college libraries. The information technology provides a wide range of services of which college libraries could avail themselves of the offers to supplement existing ones.

It is necessary to impact latest information technology skills and techniques to librarians to acquire such latest practical skills along with their professional knowledge. Otherwise it is very difficult for them to cope up with the situation in the fast changing global scenario. At times they may have to lose their ground for survival into cognizance. Library professionals should equip themselves with the new skills and practical experience in the application of information technology in the field.

POSITIVE WORKPLACE CULTURE

Development of human resources lies at the core of the knowledge-productive organizations, like libraries and universities. The process will inevitably involve participation of all employees in developing and disseminating the knowledge. The survival and success will depend on the workplace culture which includes:

- Value defined and driven
- Change orientated, flexible culture
- Effective communication

- Plans, policies, guidelines available to all staff
- Positive encouragement/support for Innovation and continuous quality improvement
- Recognition and reward systems
- Engagement with work groups and union representatives
- Management of major change procedures
- Monitoring staff satisfaction

DIFFICULTIES IN HUMAN RESOURCE DEVELOPMENT

Some of the difficulties in manpower development are mentioned below.

1. Lack of in-service training facilities.
2. Lack of enough trained personnel to maintain the database.
3. Lack of staff orientation towards automation.
4. Lack motivation.
5. Lack of knowledge and training to the users.
6. Lack of understanding between library staff.
7. Lack of awareness about the possible benefits of Information Technology.
8. Some times psychological barriers of the library staff.

Libraries are gradually shifting from manual pattern to automated pattern. Automation makes the library staff free from traditional usual routine work, and to devote time in up gradation of management and quality information services. Thus libraries should get automated not only the house keeping activities but also for providing current and relevant information to right user in the required format and in an affordable way.

The new developments are of a highly technical nature and librarians trained in the conventional library schools are not trained in know-how of the latest technology and its application to college libraries. Hence in addition to formal library education there is an urgent need to provide in-service training for library staff on the concepts, skills and technologies.

CONTINUING EDUCATION PROGRAMMES FOR LIBRARY PROFESSIONALS

Continuing education is essential for academic librarianship and can take many forms which may or may not lead to degree. Attending professional meetings and annual conferences of professional organizations, attending short-term training programmes, visiting other libraries and education institutions are some of the continuing institutions are some of the continuing education programmes for library professionals. Such continuing education programmes help them to keep abreast of all the latest information in era of knowledge explosion. The following continuing education programmes for library professionals may be offered.

NEED FOR CONTINUING EDUCATION PROGRAMMES FOR LIBRARY PROFESSIONALS

The new developments are of a highly technical nature and degree college librarians trained in the conventional library schools are not trained in the know-how of the latest technology and its application to libraries. Hence in addition to formal library education, there is an urgent need to provide in-service training to library & information science professionals on the new concepts and technologies.

The objectives of the continuing education programmes for college library staff are outlined below.

1. To help the library professionals to keep abreast of the new development in the field.
2. To make them fit and competent in the ever changing technology field.

3. To help them in improving their career prospects.
4. To help them in learning new ideas and concepts.
5. To help them to improve services.
6. To update professional knowledge.

FORMAL EDUCATION

(a) Formal Courses : Library schools will take the responsibilities in offering formal courses in handling Information Technology.

(b) Extension Courses : Summer Schools run by the Universities offer extension courses. They provide practical demonstration of different works in library.

(c) Refresher Courses : Library schools and national library associations which provide refresher courses are meant.

- To brush up all and recall acquired knowledge.
- Increasing of work efficiency.

(d) In-service Training : In-service training programmes should be organized to improve knowledge and to improve Job efficiency.

INFORMAL EDUCATION OBTAINED THROUGH:

(a) Reading, writing and editing.

(b) Attending lectures, professional meetings, conferences and workshops.

SUGGESTIONS

1. Training programmes should be conducted for library professionals to equip themselves with the application of latest information technology.

2. UGC, ICAR and Government should provide necessary funds to acquire necessary information technology equipment.
3. Directorate of High Education and Library & Information Science Schools should organize in-service training programmes regularly.
4. The parent organization should encourage library professionals in using modern technologies by providing adequate support.
5. User Orientation Programmes should be conducted for readers to utilize information technology for their needs.
6. Specialized and short term courses have to be conducted by the INFLIBNET, ICAR, Associations and Library Science Schools to library professionals to update their knowledge.

CONCLUSION

Modern information society demands for speedy information retrieval. Modern Libraries should be automated with the use of information technology application. The Academic Libraries should reorient their approach towards information technology for better prospects of development with information technology application; College Libraries can ensure increased user satisfaction. The need of the hour is to achieve better quality in collections, storage and retrieval of information. The present day environment leads to sophisticated and challenging tasks to Academic Libraries. The Academic Library Staff should improve their skills in the use of Information Technology for efficient and effective service. The government agencies, professional bodies, associations like UGC, INFLIBNET, IASLIC, ILA and other organizations related to library and information science are providing training of latest information technology by organizing conferences, seminars, workshops etc.

REFERENCES

1. Dennis R. Defa (1995). Position Analysis of Library Assistants. *Library Administration & Management* 9 (2), pp. 88–93.
2. Adedoyin, S.O. (2005). Information and Communication Technology (ICT) Literacy among the Staff of Nigerian University Libraries. *Library Review* 54 (4), pp. 457-266.
3. Hayes, Helen (1998). New Missions of academic libraries in the 21st Century. (Beijing Conference- Report). Council of Australian University Librarians.
4. Sridhar, M.S (1999). Skill requirements of LIS Professionals in the new e-world. *Library Science with a Slant to Documentation and Information Studies* 36 (3).
5. Fitzgibbons, Megan Anne (2008) Human Resources Management in the Development of Information Commons in Academic Libraries. *Library Student Journal.*
6. Christopher, C. (2003). Empowering your library: a guide to improving service, productivity, & participation. Chicago: American Library Association.
7. Pomerantz, J. (2003). Integrating Digital Reference Service into the Digital Library Environment. In R. D. Lankes, S. Nicholson & A. Goodrum (Eds.), The Digital Reference Research Agenda (pp. 23-47). Chicago: Association of College and Research Libraries.
8. Godoa, K. (2007). Access and the Changing Trends in Libraries. New Delhi: Adhyayan.
9. Jain, S.L. and Gupta, D.K. (1996). TQM in Library and Information Services. *Annals of Library Science and Documentation.*
10. Devi, P. and Singh, S. (2006). Role of UGC in Manpower Development in the Field of Library and Information Science in India. *Annals of Library and Information Studies.*

OPEN ACCESS INITIATIVES AND IMPLICATIONS

INTRODUCTION

Open Access is a *buzz* word in present day knowledge society. The growth of the open access movement is partially in response to the enormous costs of many scholarly information sources. With conventional journals publication methods it is uncommon for an institution to have to pay for an article twice. First they pay the scholars to produce the work and then the institutions library pays to purchase the same literature back from the publisher.

The present day society is largely depends on information to sustain in the society, it is also essential to perform any task being it a personal, professional, organizational and what not, whatever we do in this society is depends upon the information we possess. At this juncture we must make a note of information rich and information poor. Open Access can bridge the gap between the information rich and information poor through the availability and accessibility of information freely, easily without any restrictions to all. Now a days the library and information centres are making several efforts to accumulate all the open access resources at one point and directly providing hyper-links to them in the library website to facilitate the user in browsing those resources.

OPEN ACCESS

An Open Access or Open Source will comply the following:

- The acceptance of the free re-distribution;
- The programme must include the source code;
- The license must allow modifications and derived works;
- The integrity of the author's source code must be maintained.

WHY OPEN ACCESS

With escalating printing prices and dwindling budgets in all sectors, library and information centres are forced to look for the alternative solutions which provides access to scholarly resources. It is the ability to find electronic information that is only accessible at a meager or no cost is the main driving force of Open Access movement.

Open Access resources and documents can be located through World Wide Web search engines such as Google, Alta Vista, and Google Scholar etc. For optimal visibility and scientific impact, documents must also be findable via reference services which are frequently used by the scholars and scientists from their area of interest. The global linking of archives and the fact that they provide value-added services viz. citation count, reference linking, RSS feeds etc. can further the usability, impact and visibility of open access resources.

TYPES OF OPEN ACCESS

There are mainly two types of open access resources published by the scholars and publishers

- **Green Route:** The author can self-archive at the time of submission of the publication. Whether the publication is grey literature (generally internal non-peer-reviewed), peer-reviewed journal publication, peer-reviewed conference proceedings paper or a monograph. This kind of initiative is known as green route.
- **Gold Route:** The author or author institution can pay some fee to the publisher at the time of publication, the publisher thereafter making the material available 'free' at the point of access. This process is called as gold route.

In simple terms a researcher can place a copy of each article in an Open Access repository or can publish articles in Open Access journals. In addition, a researcher may place a copy of each article on a personal or departmental website. These entire efforts route to Open Access ensure that far more users can access such articles than if they were hidden away in Closed Access journals. The first two constitute much more systematic and organized approaches than the third and optimize the chance of other researchers locating and reading articles.

OPEN ACCESS INITIATIVES

The open access movement has taken momentum by encouraging more and more publishers mostly learned societies to join the cause for providing free access to the scholarly literature. Some of the important initiatives are:

- **Budapest Open Access Initiatives (BOAI):** BOAI has been signed by a increasing number of Researchers, Universities, Laboratories, Libraries, Foundations, Journal Publishers, Learned Societies and Scholars from around the globe. It recommends using two complementary strategies viz. self-archiving in institutional/subject repositories and open access journals.
- **Public Libray of Science (PLoS):** It is grass roots initiative signed by over 30,000 scientists across the world to encourage publishers to deposit their journals in central archives like PubMed Central, within six months of publication. Later due to modest response from publishers, PlOs became publisher to launch its first open access journals PLoS Biology.
- **Scholarly Publishing and Academic Resources Coalitions (SPARC):** SPARC is actively promoting both open access journals and the development of institutional repositories. It has a number of open access partners including; Algebraic and Geometric Topology, Documenta Mathmatica, Scholarship, Geometry & Topology, Economics bulletin, etc. It also brings out lower-cost, directly competitive journals as an alternative for academic disciplines formerly dependent on high-priced journals.
- **Directory of Open Access Journal (DOAJ)**: DOAJ is a gateway that aims to link all open access scientific journals that use an appropriate quality control system. It is managed by LLND University and supported by Soros Foundation, and currently

access to over 5000 electronic journals in all academic disciplines in different languages.

- **Access to Global Online Research in Agriculture (AGORA):** Food and Agriculture Organization, United Nations together have initiated it. The AGORA is tie up with major publishers and provides online free access to nearly 900 commercially published journals in the fields of Food, Agriculture, Environment Sciences and allied disciplines in 113 developing countries all over the world.

- **Agricola:** This initiative has been materialized with the efforts of United States Dept. of Agriculture. It provides a comprehensive source of bibliographic citations covering U.S. Agricultural and Life sciences information. Agricola contains over 35 lakhs citations to scholarly journal articles, monographs, theses, patents, software, audio-visual materials, and technical reports related to all aspects of agriculture. Coverage is from 1970 to the present, with monthly updates. It also provides some citations in French.

Another important initiative is Bethesda Statement on Open Access Publishing. It continued to promote a gradual transition to open access publishing, although within the biochemical community. These are some of the important initiatives for promoting open access. Although neither of these early initiatives outlined the impact or implications of the open access movement on developing countries, there is the philosophical imperative intrinsic that the idea of free, equal and open access would apply to all communities and countries.

OPEN ACCESS IMPLICATIONS

- Access to un-restricted scholarly information resources.
- Open Data
- Self Archiving in repository
- Providing Open Access to the articles will have greater impact on citation advantage.

- Open Access institutional repositories benefit institutions by supporting information management
- Open Access publication allows the authors/creators to retain the copyright.
- Open Access helps in detecting plagiarism more easily.
- Open Access archives are cost effective.
- Open Access does not affect peer-review.

OPEN ACCESS – MISCONCEPTION

In the context of commercial world, one cannot expect anything for free of cost. Everything which possesses some value has been commercialized. There is one misconception that, if anything is available free many thinks that it has no value. It is very obvious and misconceptions regarding open access. It is not self-publishing, nor a way to bypass peer-review and formal publication, nor is it a kind of second category, cut-price publishing route. But it is simply the means to make research results freely available online to the whole research community for further improvement and ramification of phenomenon.

CONCLUSION

Open Access is a gift to the research community where they can search the information at their desktop/laptop sitting at their places. It does not require any User ID, Password and I.P. Address. On the other hand the libraries are benefited a lot from the Open Access initiatives which saves financial resources as well it helps to improve the value in the user community. However, Open Access progress is comparatively slow in developing countries like India, Latin America and Africa are realized the importance of Open Access and strives to improve the distribution of knowledge through Open Access Publishing. In our country there are several workshops and seminary being conducted the Indian Academy of Science to create awareness of Open Access. The Open Access initiatives will successful only when huge number of users will make use of it and in return they also support and contribute to the growth and development of Open Access.

REFERENCES

1. Canada, Dayna. (2009) Open Access and Developing Countries. *Open and Libraries Class Journal* 1 (2).
2. Christian, G.E. (2006) Open Access Initiative and the Developing World. *African Journal of Library Archives & Information Science* 18(1), pp. 1-9.
3. Serials Review Special Issue. (2004) *Open Access* 30(4).
4. Nazim, Mohd. And Maya Devi (2008). Open Access Journal and Institutional Repositories Practical Needs and Present Trends in India. *Annals of Library Science and Documentation* 55(3), pp. 27-34.
5. Kwan, J. (2003). What is Open Access and Why Should You Care? Latitudes, 12 (3).
6. Pinfield, Stephen. (2007) Libraries and open Access: the implications of Open Access Publishing and Dissemination for Libraries in Higher Education Institutions.

WEB SOURCES

1. www.plos.org
2. www.ncbi.nlm.nih.gov/entrez
3. http://www.soros.org/openaccess/software/
4. http://www.doaj.org
5. http://www.aginternetwork.org/en/journals.php
6. http://agricola.nal.usda.gov
7. Bethesda Statement on Open Access Publishing. (2003) http://www.earlham.edu/~peters/fos/bethesda.htm

8. Berlin Declaration on Open Access to Knowledge. (2002) http://www.earlham.edu/~peters/fos/bethesda.htm
9. http://www.doaj.org
10. http://www.earlham.edu/~peters/fos/bethesda.htm
11. http://www.nnlm.gov/psr/lat/v12n3/open_access.html
12. http://eprints.nottingham.ac.uk/697/>

SCHOLARLY INFORMATION AND COMMUNICATION IN THE WEB ENVIRONMENT AT ANGR AGRICULTURAL UNIVERSITY

INTRODUCTION

Information Technology is rapidly changing the whole world creating new challenges and opportunities. ICT has tremendous impact on library operation, resources, services, staff and users. In recent years, the Internet and especially the web has brought information communication revolution enabling us to access, retrieve and send information anywhere instantaneously and inexpensively. The web has transformed the way of libraries functioning and providing library information services to the user community. The web contains huge volume of global information on wide range of subjects. It has made a profound contribution to all the spheres of modern life. Web appears to be almost totally free. Information access is now so freely available through web. The web based services offer high quality, easy to use methods for finding global information at the users desktop.

The web has exposed the limitations of library information service available at physical building, with limited opening hours. More fundamentally, Internet users simply expect to be able to access any information they want, from anywhere in the world, at any time.

SCHOLARLY COMMUNICATIONS

Scholarly communication is an umbrella term used to describe the process of academics, scholars and researchers sharing and publishing their research findings so that they are available to the wider academic community. The term 'Scholarly Communications' has been in common usage at least since the mid-1970s, in recent years there has been widespread belief that the traditional system for disseminating knowledge.

MEANING

Scholarly communication is the creation, transformation, dissemination and preservation of knowledge related to teaching, research and scholarly endeavors. Among the many scholarly communications issues include author rights, the economics of scholarly resources, new models of publishing including open access, institutional repositories, rights and access to federally funded research, and preservation of intellectual assets.

Scholarly Communication is seen as a crucial part of research and researchers. The research output of an institution is often judged by their academic output and list of publications. A researcher's publication will help to create his reputation within his discipline. The digital revolution has resulted in drastically simpler, timelier and access to a wide range of information.

ANGR AGRICULTURAL UNIVERSITY LIBRARY SYSTEM

With a view to cater the library and information needs of teachers, students as well as research and extension personnel, the library system was reorganized in 1980, forming Central Library at Hyderabad and Regional Libraries at Tirupati and Bapatla and College Libraries at other campuses. All the Libraries are computerized during the AHRD Project. However with the NATP support during 2001-2005, the Central Library has been completely automated and a large collection of theses has been digitized.

ANGRAU - Library System

Central Library	Rajendranagar, Hyderabad
Branch Library	Extension Education Institute, Rajendranagar
Regional Libraries	S.V. Agricultural College, Tirupati Agricultural College, Bapatla

Contd....

College Libraries	College of Home Science, Hyderabad College of Food Science & Technology, Bapatla Agricultural Engineering College, Bapatla Agricultural College, Aswaraopet Agricultural College, Naira Agricultural College, Mahanandi Agricultural College, Jagtial College of Agriculture, Rajahmundry College of Food Science & Technology, Pulivendula College of Agricultural Engineering, Madakasira

ROLE OF ICAR IN PROMOTING RESEARCH AND COMMUNICATION

Indian Council of Agricultural Research (ICAR), New Delhi, India is an autonomous organization under the Department of Agricultural Research and Education, Ministry of Agriculture, Government of India. The Council is the apex body for co-ordinating, guiding and managing research and education in agriculture including horticulture, fisheries and animal sciences in the entire country. With over 90 ICAR institutes and 45 agricultural universities spread across the country this is one of the largest national agricultural systems in the world. The ICAR has initiated to strengthen the libraries attached to ICAR institutions and SAUs. The following programmes are mainly aimed at capacity building of agricultural libraries.

- **CeRA**
- **KrishiPrabha - Indian Agricultural Doctoral Dissertations Repository**
- **e-GRANTH**

CeRA (Consortium for e-Resources in Agriculture)

Consortium for e-Resources in Agriculture has been established under the NAIP (National Agricultural Innovation Project) for providing online access to e-journals and resources in over 120 ICAR libraries.

Objectives

- To develop the existing R & D information resource base of ICAR institutes/ Universities etc., comparable to that existing in world's leading institutions/ organizations.
- To subscribe e-journals and create an e-access culture among scientists/teachers in ICAR institutes/agricultural universities.
- To develop a Science Citation Index (SCI) facility at IARI for evaluation of scientific publications.
- To assess the impact of CeRA on the level of research publications measured through SCI.

CeRA Participating agencies

Sl.No.	Organization type	No.
1.	Deemed Universities	5
2.	ICAR Headquarters and Krishi Anusandhan Bhawans	3
3.	National Bureaus	5
4.	Research Institutes	42
5.	National Research Centres	21
6.	Project Directorates	9
7.	State Agricultural Universities	38
	Total	**123**

E-Resources available under CeRA

Sl.No.	Name of the Publisher	No. of Journals
1	Springer Verlag	70
2	Annual Reviews	22
3	CSIRO Publishers	08
4	Open J- Gates	613
5	Science Directory	363
6	Taylor & Francis	1079
7	Indianjournals	131
8	Elsevier	358
9	Oxford Journals	30
10	American Society of Agronomy	08
Total		**2682**

KrishiPrabha

KrishiPrabha is a full-text database of Indian Agricultural Doctoral Dissertations submitted by research scholars to the 45 State/Deemed Agricultural Universities during the period from 1.1.2000 to 31.12.2007. This database, listing about 7500 Doctoral Dissertations with a full text of about 6000 Dissertations.

e-Granth

Strengthening of Digital Library and Information Management under NARS (e-GRANTH) is a subproject under NAIP. e-Granth is one of the prominent research projects pertaining to library which aimed at to strengthen the delivery of information with collaboration among the 12 major agricultural libraries in the country.

OBJECTIVES

To create Online Public Access Catalog (OPAC) under "Indian Agricultural Research Group Catalogue" of all 12 library resources with Online Computer Library Center (OCLC) partnership.

- To digitize important institutional repositories (limited to IARI, New Delhi; IVRI, Izatnagar; ANGRAU, Hyderabad and UAS, Bangalore) including rare books and old journals and make them open access under NARS.
- To strengthen capacity building for library and information management system (open to all libraries of NARS).

AgriCat

AgriCat is the outcome of e-Granth project. It provides digital access to library resources of 12 agricultural research institutes and agricultural universities which include OPAC, important institutional repositories, rare books and old journals of importance. Presently it holds nearly 5 lakh bibliographical items of consortium partner libraries.

ANGRAU RESEARCH OUTPUT

Sl.No.	Publication Year	Total Publications
1.	2005-2006	270
2.	2006-2007	262
3.	2007-2008	288
4.	2008-2009	293
5.	2009-2010	302

From the above table it is seen that there is a visible growth in the total number of research papers published by ANGRAU scientists, which is indicative of the availability of scholarly information provided by the library.

ANGRAU Library Publications

ANGRAU Central Library has started the following publications for communicating Research Information to Agricultural Scientists, Teachers, Extension Specialists and students.

- Current Agricultural Titles (CAT) – Print and CD form

- ANGRAU Library Bulletin
- ANGRAU Theses Abstract

ONLINE SERVICES PROVIDED IN ANGRAU LIBRARY SYSTEM

ANGRAU provides the scholarly information to its readers through the following online sources.

- CAB Online Abstract Services
- J-Gate Online Services
- EBSCO : E-Journals Databases
- Indianstat.com
- Indianjournals.com
- Globalperiodicals.com
- Consortium for e-Resources in Agriculture (CeRA)
    - Springer Verlag
    - Annual Reviews
    - CSIRO Publishers
    - Open J- Gates
    - Science Directory
    - Taylor & Francis
    - Indianjournals
    - Elsevier
    - Oxford Journals
    - American Society of Agronomy
- Krishi Prabha - Indian Agricultural Doctoral Dissertations Repository

PROMOTING SCHOLARLY COMMUNICATION - ROLE OF ANGRAU LIBRARIES

"Think globally and act locally" the phrase is very much suitable for present day libraries. ANGR Agricultural University is always standing in forefront in providing latest and timely information to the agriculture user community. Some of them are

- Acquire latest/current information resources
- Build appropriate combination of both print and electronic information resources
- Provide information support for research
- Computerization and networking of all library operations, resources and services
- Provide access to on-line journals on the web
- Assessment of user needs by interacting with users
- Provide guidance and training in the usage of digital resources/e-resources
- Providing a gateway access to the content of local and global resources and services
- Provide responsive products and services to meet the user needs and expectations
- Create Institutional Repository of ANGRAU research output
- Capacity building of libraries & library staff in adopting latest techniques

CONCLUSION

In the increasingly complex world of scholarly communication and the varied technologies, tools, and services available to libraries; including institutional repositories, open access initiatives, self-publishing/ e-publishing models, librarians are in a position to envision, shape, and articulate their future role in scholarly communication. The potential utility of the scenarios for individual libraries depends on the application of the possible, plausible, or inspirational futures question posited by one participant, while taking into consideration the local environment in which the library is set. Day to day Librarian faces many challenges in digital era to maintain the library up to the mark. The collaboration and consortia is critical that libraries continue to work within their institution along with stakeholders and with partner institutions to create new scholarly communication structure and cultivate a prominent role in the new model.

REFERENCES

1. ANGRAU Annual Reports
2. Chad, Ken and Miller, Miller (2005) Do Libraries Matter? The rise of Library 2.0. *A white paper*. P. 1-11.
3. Kresh, Diane ed. (2007). The whole digital library handbook. Council on Library and Information Resources.
4. Carpenter, Maria; Graybill, Jolie and Piorun, Mary. Envisioning the Library's Role in Scholarly Communication in the Year 2025.
5. Ekman, Richard and Quandt, Richard E (1999) Technology and scholarly communication University of California Press.

WEB SOURCES

1. http://en.wikipedia.org/
2. http://hillagric.ernet.in
3. http://www.angrau.net
4. http://www.cera.jccc.in
5. http://www.egranth.ac.in/
6. www.hau.ernet.in
7. www.icar.org.in

MANPOWER DEVELOPMENT IN INFORMATION TECHNOLOGY ENVIRONMENT IN AGRICULTURAL LIBRARIES

The progress of research in agricultural fields and improvement of production, essentially require timely supply of information to agricultural scientists. It is not possible to achieve self-sufficiency in food production without adequate information supply to agricultural scientists.

Information Technology now impinges on virtually every aspect of society and has developed dramatically at both extremes of its use. For all areas of science, information technology is now of immense importance, both as research tool and as a medium of facilitating the information transfer process.

It is also not difficult to envisage all the components of information technology to be used in agricultural research and extension institutions in developing country like India. While there are undoubtedly barriers for developing countries, not the least of which will be the costs, the potential for using advanced information technology is undeniable, and undoubtedly opens up new vistas for the transfer of scientific information through projects like Agricultural Research Information System of ICAR, which is a boon to Agricultural Research Community.

It is essential to build the requisite information infrastructure for agricultural scientists within the state and nation. This is particularly relevant in the context of "food for all". In pursuit of scientific excellence, agricultural information service facility keeps track of the latest scientific, agricultural and technological advances all over the world.

OBJECTIVES OF INFORMATION TECHNOLOGY

The objectives of information technology are to provide better means of transmission of data or messages in the form of written or printed records, electric, audio or video signals by using wires, cables and telecommunication techniques. Information technology plays a vital role in information handling due to developments such as reduction in computing time, capabilities of resource sharing, economic storage capacities of files on CDs, use of TV as ready made information screen, telecommunication and satellite communication facilities, etc.

The applications of information technology in agricultural libraries and information centers can be categorized into the following:

- Supporting technical functions associated with technical processing and circulation work.
- Supporting information storage, retrieval and dissemination system.
- Supporting management information services for libraries.
- Supporting professional and orientation courses for practicing libraries, continuing education
- Programme for teachers of library and information science, correspondence studies and library extension services, etc.

NEED OF INFORMATION TECHNOLOGY IN AGRICULTURAL LIBRARIES

Information Technology can be effectively applied to agricultural library services in view of the following advantages and necessities:

- Information Explosion
- Rise in the cost of print media
- Change in the profile of library users.
- Space problems
- Change in the mode of services i.e. from collection based services to information based services.

- ❖ Knowledge through Internet and global information superhighway etc.
- ❖ Availability of Scholarships and fellowships.
- ❖ Information about Career guidance.

FACTS OF INFORMATION TECHNOLOGY

Various facets of Information Technology viz., media, tools and gadgets are growing and changing very fast. It has drastic influence on the environment of agricultural libraries and of their working. Some of these are:

- ➢ Audio and video technology ~ Micrographic technologies
- ➢ Optical technologies
- ➢ CD-ROM technologies
- ➢ Video disk technology
- ➢ Satellite technology
- ➢ Facsimile technology
- ➢ Video and Teletext technology
- ➢ Hypertext and Hypermedia technology
- ➢ Electronic Mail and Message System (EMMS) technology
- ➢ Voice mail technology
- ➢ Bar code technology

ADVANTAGES OF INFORMATION TECHNOLOGY IN AGRICULTURAL LIBRARIES

Some of the advantages of using Information Technology in Agricultural Libraries are:

- ❖ Ability for generation of new services

- Accommodation of increased workload
- Facility co-operation
- Achievement of greater efficiency

IMPACT OF INFORMATION TECHNOLOGY ON AGRICULTURAL LIBRARY SERVICES

There are several means of modern technology, which brought in many services to agricultural libraries so as to speed up their activities; they include

- Telecommunication technology
- CD-ROM technology
- On-line retrieval services
- Library networks
- E-Mail
- Internet

New technology now available will help to remove barriers of distance of time and drudgery of repeated manual efforts in library routines. The advancement in technology will continue to improve upon the effectiveness of the libraries and indispensable for handling messages between libraries and library patrons.

NEED OF MANPOWER DEVELOPMENT IN INFORMATION TECHNOLOGY ENVIRONMENT IN AGRICULTURAL LIBRARIES

The application of modern technology allow to access and store information quickly availability of computers has gradually changed the agricultural libraries. The information technology a wide range of services of which libraries could avail themselves offer to supply existing ones.

Some of the areas to which information technology could be applied in day-to activities of the libraries are; acquisition, cataloguing, circulation, indexing and abstracting articles etc. Therefore, it is necessary to impart latest information technology skills, techniques to agricultural librarians to acquire such latest practical skills along with professional knowledge. it is very difficult for them to cope with the situation in the changing global scenario and at times they may have to loose their ground for survival also. Taking the present day situation into cognizance, agricultural library professionals should inform themselves with the new skills and practical experience in the application of information technology in the field.

DIFFICULTIES IN DEVELOPING MANPOWER

1. Shortage of information professionals' skills in using information technology.
2. Lack of in- service training facilities.
3. Lack of enough trained personnel to maintain the database.
4. Lack of staff orientation towards computerization.
5. Lack of motivation.
6. Lack of knowledge and training to the users.
7. Lack of understanding between library staff.
8. Lack of awareness about the possible benefits of Information Technology.
9. Some times psychological barriers of the library staff.

SUGGESTIONS

1. Training programmes should be conducted for staff to know the latest information technology.

2. ICAR should provide necessary finance to State Agricultural Universities and Agric Research Institutes to acquire necessary Information Technology equipment.
3. ICAR should organize in-service training programmes.
4. User orientation programmes should be conducted for readers to use information technology.
5. The Agricultural libraries staff should be encouraged in using modern technologies.

INFORMATION SERVICES IN AN ELECTRONIC ERA VIDE ADOPTION OF NEWER TECHNOLOGIES ANTICIPATION VISION 2020 OF ANDHRA PRADESH

INTRODUCTION

Life long learning is becoming essential for everyone otherwise one is becoming irrelevant, unwanted and therefore stands a chance of being left out. Libraries have been down the ages, store housed of knowledge. Until recently, that store was in the form of books. One has to be a member of library, physically go to it and borrow books. This involves the obvious limitations of time, conveniences and also expense. The digitization of information and its storage in solid state memories of computers and the ability to access these stores of digitised information over the global telecommunication networks from offices and homes and now a days even while on the move, is profoundly changing the concept of libraries, storage of books and the limitations on space and country and time. Every work of an author, artist and philosopher and every piece of information can be put on the world wide web (WWW). Libraries can be totally web based and ultimately free of books as we know. These can be accessed over the global telephone network from communication terminals on their parson. So far it is through PC's that the Web could be accessed and information seen on the screen and then printed if wanted.

NEED FOR ADOPTION OF NEWER TECHNOLOGIES

For scientific and economic development: The developing countries have to open up under the impact of liberalizations and globalisation. After five decades of independence, towards the close of the century, the dependence of developing countries on the western products more so when consumerism is on the rise, has rather increased. In this age , as most of the information products like books, journals CD-ROMs, data bases, electronic journals etc., are being designed in the developing countries so knowledge. In such a situation the need to adopt the new technology will become imperative sooner or later, in order to ensure accessibility to the new formats in which the information of the future will be packaged. Meanwhile technological developments during the last few years have helped to decrease the hardware costs to affordable level. Although there remain the software problems, but these two have been solved to some extent.

TECHNOLOGH VS INNOVATION

The major objective of any University/Institution, Organisation is to train man power through imparting education conducting location specific research and organize extension activities for the benefits of the personnel of the line departments of the Government, NGO's, farmers and others. At every stage of these activities, availability of information, accessibility of information and affordability of information to the user community is the major concern. The user community is a heterogeneous group of students, teachers, researchers, expansionists, Businessmen. Industialists, subject matter specialists, planners and policy makers, they all need information at a right time at a right place, in a right from. In providing information the library and information centres need to engage in strategic visioning in exploring the technological challenges. They need to evolve to new organizational forms which are being enabled by anytime, any place information the reduction of barriers of time, and both geographical and organizational distances.

Innovation has a historical background. We know how successive generations in the past have introduced innovations in the field of libraries. From chained books to closed access; from closed access to open access; from open access to extension services now we find the information super highway and networks.

Today books and reading no longer possess a unique informative power that once was theirs. Various media such as films tapes, videotapes, records, telex, fax, DTP, E-Mail, Television and computers now serve to store and transmit information. We know the innovations in the form of online information systems under these circumstances, it is highly difficult to predict what will happen tomorrow. After all, one innovation aims at acting as a catalytic agent between the information user and the creator of information

Innovation refers to doing new things and new way. Extrapolation can be said as using technology to do what libraries are doing now in a faster, cheaper, and better way. Examples include online card catalog. Enriching the catalog with line links to bitmapped, facsimiles of printed pages, hybrid libraries, digital library, federations hubs of community network providers etc.

PERSONNAL COMMUNICTION SYSTEMS

The days of place centered libraries are counted. Personal computerized libraries are in the offing. Personal communication systems are in great use. Most network traffic today is people talking to one another. In such a set up, the libraries trained mainly to practice at place centered libraries of physical objects will be of little use unless they are moulded to fit into the changing situation. The library buildings, the security systems and what not every thing in the libraries will have a see change by 2020. The onerous of adjustment rests with the mindset of authorities that govern these institutes.

It is commonly agreed that if there is any science that will dominate the 21st century, it is the information science. If there is one industry in which India can emerge as a world leader, it is the information technology. One more dimension can be added to this: if at all there is any science which can help in the development of any nation culturally and scientifically it is the library and information science.

VISION 2020 AND ANDHRA PRADESH

No government can afford to ignore modernization of these libraries to give to all of them internet connection and computerize every one of them. Every new book that is being published should be digitized and committed to digital memories. Many of our libraries, especially in the universities and important educational institutions must create their own web sites and networks so that the total number of books that are in the state can be in a pool and any member of a library anywhere in the state would be able to browse first and then get the book of information he wants. This is a great job and unless the vision 2020 of the state provides for the realization of this prospect, it would not be fulfilled.

As a result of the implantation of vision 2020 all Government offices up to mandal level will be interconnected on broad based, high speed digital, optical, fiber network over which information ranging from data to video images could be transmitted. All schools, colleges, other educational institutions and hospitals will be equipped with personal computers. Every village will have a PC just like the present STD/ISD public telephone booths. Those who cannot afford telephones can have voide mail box so that no home remains 'unmassaged'.

Such extensive development of information technology will speed up literacy, enhance the quality of education, help citizens to be information rich, enable them to participate in and influence governance, effect economy and efficiency in government functions and fulfill people's demand for a transparent government which commits to furnish information on demand to any citizen through PC-based information Kiosks. The vision envisages that 85 percent of service sector people will work with PC's: 90 per cent of bank transactions would done without the need to go to bank; 65 percent of homes would have cable TV and telephone/voice mail boxes.

CONSLUSION

Finally, it is to conclude that we are opt to make best use of newer technologies from time so as to construct a track anticipating vision 2020 of Andhra Pradesh. Though is a dream for the present, we can make use of latest information services to expect everything in reality. Information technology activates its strength by exploring the technological challenges. Libraries and Information centres are exceptional means of motivating the people for the cause.

REFERENCES

1. Chowdhary, T.H. (1997). Information technology vision 2020, Govt. of Andhra Pradesh. Draft .
2. Sudhakar Reddy, S (1997). Vision 2020-Mission Statement. Economic Times (11th Sept).
3. Jayadav, K (1998). Smart government with vision for 2020. Express Computer (17th August).

USEFUL INFORMATION REQUIREMENTS OF AGRICULTURAL SCIENTISTS : AN OPINION SURVEY IN ANDHRA PRADESH

INTRODUCTION

The progress of research in Agricultural fields and improvement of production, essentially require timely supply of information to agricultural scientists. It is not possible to achieve self sufficiency in food production without adequate information supply to agricultural scientists.

The concentration of effort is probably efficacious in informing agricultural scientists of improved methods in agriculture. It also indicates an unfortunate parochialism effecting both the agricultural scientists and the information handling industry. To promote an effective working relationship between research workers who generate new agricultural knowledge, farmers who stand in need of such knowledge, and Information system acts as 'middlemen' in a position to deliver such knowledge to the required points of distribution. The sources of agricultural information are varied in kind and widely scattered in location with an equally widespread distribution. In the same way agricultural information users are also equally scattered. To bridge these two, we need agricultural information networks.

OBJECTIVES OF THE STUDY

1. To know the opinion of agricultural scientists regarding adequacy of books and periodicals.
2. To know what extent the agricultural libraries are meeting the requirements of agricultural scientists.
3. To know the opinion of agricultural scientists about adequacy of agricultural information sources.

SAMPLING TECHNIQUE

To collect data from scientists, the following sampling design has been adopted.

All the users of the libraries of agricultural colleges and research stations in Andhra Pradesh Agricultural University (Now changed the name as Acharya N.G. Ranga Agricultural University). It is estimated that nearly 850 agricultural scientists are using the agricultural libraries in Andhra Pradesh.

Having identified the population, the researcher then selected a sample by adoption of a stratified random sampling. Primarily the population (Scientists) were stratified into faculty members and researchers. Then the faculty members and researchers were further stratified according to their specialization. From the group of faculty members as distributed according to their specialization, a sample of 100 were drawn at random and from among the group of researchers as distributed according to their specialization a sample of 150 were drawn at random.

LIMITATIONS

Not all agricultural scientists were taken for this study. Scientists belonging to other than Andhra Pradesh Agricultural University were not explored due to paucity of time and money.

DATA ANALYSIS

The primary data collected through questionare was analysed by using statistical technique Chi-square test.

Table 1 Respondents by opinion regarding Adequacy of Book in the Library

SI. No	Opinion	No. of Respondents		Total
		Teaching	Research	
1	Inadequate	12 (12)	102 (68)	114 (46)
2	Adequate	74 (74)	35 (23)	109 (43)
3	Fairly adequate	14 (14)	13 (9)	27 (11)
	Total Chi square value	100 (100)	150 (100)	250 (100) 77.8*

* Significant at 1% level.

From the Table 1 it is found that 74% of teachers and 23% of researchers expressed that the books in the agricultural libraries were adequate, 12% of teachers and 68% of researchers expressed that books are inadequate. Only 14% of teachers and 9% of researchers are feeling that the books are fairly adequate. The majority (46%) of respondents have expressed that the books in the agricultural libraries are inadequate.

An attempt is also made to know whether there exists a significant association between the classification of scientists as teachers and researchers, and their opinions with regard to adequacy of book in the library. The null hypothesis is formed that the opinions are independent of type of scientists and tested with a test statistical Chi-square. Since the calculated value of chi-square is much greater than the table value at all levels of significance, the null hypothesis is rejected. Therefore, it can be inferred that the opinions are not independent of the type of scientists. In other words, the opinions with regard to adequacy of books vary depending upon whether the scientist is a teacher or a researcher.

Table 2 Respondents by opinion Regarding Adequacy of periodicals in the Library

Sl. No	Opinion	No. of Respondents		Total
		Teaching	Research	
1	Inadequate	23 (23)	124 (83)	147 (159)
2	Adequate	56 (56)	21 (14)	77 (31)
3	Fairly adequate	21 (21)	5 (3)	26 (10)
	Total	100 (100)	150 (100)	250 (100)
	Chi square value			88.7*

* Significant at 1% level. ,

Majority (83%) of researchers have expressed that the periodicals are inadequate and majority (56%) of teachers have expressed that the periodicals are adequate. 59% of respondents have expressed that the periodical collection inadequate, 31% of respondents adequate and 10% fairly adequate. It is clear that periodical is not adequate.

Further to establish statistical relationship between the .opinions and classification of respondents, the chi-square test is used. The null hypothesis framed in this context is that the opinions expressed by the scientists are independent of type of respondents namely teaching and research. As the calculated value of the chi-square statistics are much higher than the table value at 1% level significance, we reject the null hypothesis. Thus we conclude that opinion expressed by the respondents about the periodicals is depending upon the type of respondents.

Table 3 Respondents by opinion regarding Adequacy of Books in the Library

SI.No	Extent	No.of Respondents		Total
		Teaching	Research	
1.	0-20%	8 (8)	15 (15)	23 (9)
2.	20-40%	18 (18)	72 (48)	90 (36)
3.	40-60%	46 (46)	55 (37)	101 (40)
4.	60-80%	21 (21)	8 (5)	29 (12)
5.	80-100%	7 (7)	-----	7 (3)
	Total Chi square value	100 (100)	150 (100)	250 (100) 39.76*

Significant at 1% level.

Table 3 shows the respondents opinions regarding the extent of the library in meeting their requirements. The highest number of respondents (40%) expressed that 40-60% are only meeting their requirements. Only 7% of teachers have expressed that the libraries are meeting 100% of their requirements. And no researcher have responded that 80% to 100% of meeting their requirements. Highest number of teachers are responded to 40% to 60% and researchers are responded to 20% to 40%.

The above analysis reveals that the agricultural libraries of Agricultural University are not meeting the requirements of the users. Hence, it is suggested to libraries to join the networking system to meet the requirements of the readers.

To test statistically, a null hypothesis is framed that the extent the library is meeting their requirements and the classification of the respondents are independent. Since the table value is greater than the calculated value Chi-square, the null hypothesis is rejected. Hence it may be stated that there exists an associate between the opinion about the library in meeting the requirements and the classification of the respondents.

Table 4 Opinion of Respondents Regarding Adequacy of Agricultural Information Sources

SI.No	Opinion	No.of Respondents		Total
		Teaching	Research	
1	Sufficient	14 (14)	16 (11)	30 (12)
2	Moderately Sufficient	23 (23)	42 (28)	65 (26)
3	Insuffient	63 (63)	92 (61)	155 (62)
	Total	100 (100)	150 (100)	250 (100)
	Chi square value			1.159*

* Not Significant at 5% level

Table 4 shows the respondents opinion of level of adequacy of agricultural information sources. 62% respondents have expressed insufficient, 26% moderately sufficient and 12% of respondents have mentioned adequacy of agricultural information sources. Majority of teachers and researchers expressed that the agricultural information sources are insufficient. The analysis reveals that the agricultural information sources are insufficient. This may be due to the less number of sources available in the agricultural libraries.

A null hypothesis is framed and its validity is tested statistically by showing chi-square test statistic as the chi-square calculated value is less than the table value at 5% level of significance, we accept the null hypothesis which is stated as follows.

Ho = The opinion of the respondents about the adequacy of agricultural information sources independent of the type of respondents.

FINDINGS

The following findings were drawn from the analysis of data.

1. Majority of the agricultural scientists have expressed that the books and periodicals in agricultural libraries are inadequate.
2. In the opinion survey, the agricultural scientists have expressed that the periodicals are inadequate. Also the opinion about the periodicals will depend on type of respondents.
3. The analysis shows that the agricultural information requirements of the users are being satisfied in different levels. It is but natural; the qualitative and quantitative information varies from time to time. It all depends on the nature and intensity of the agricultural research problem that requires information.
4. Majority of the agricultural scientists have expressed that the agricultural information sources are insufficient to meet their information requirements.

SUGGESTIONS

Based on the conclusions, the following suggestions were made,

1. To overcome the financial crisis being faced by the agricultural libraries it is suggested to encourage in mobilizing their own financial resources either by marketing information or information consultation services.
2. The Indian Council for Agricultural Research should establish agricultural resource centres/informatics in specialized disciplines to meet the individual information requirements of the agricultural scientists.
3. Agricultural Information materials available in all Agricultural colleges, libraries and agricultural research centres are to be accessible to all the Agricultural scientists residing in Andhra

Pradesh state irrespective of their belonging to any particular institute.

4. There is a need for wider circulation of agricultural publications and also wider circulation of Information about new arrivals. This has to be done by a central agency incorporating all materials at one place and this catalogue should be made available to agricultural scientists, extension workers, planners and farmers.
5. The collection of documents in the agricultural libraries of Andhra Pradesh should be improved in order to meet the needs of the agricultural scientists.
6. Duplication of document sources are to be avoided to save Money.
7. ICAR should undertake the responsibility in planning National Agricultural Information Networks for all agricultural libraries in India and it should provide agricultural information to the scientists. They should also take the responsibility of supplying hard copies of books etc at their arrivals to the agricultural scientists.

CONCLUSION

Information seeking is a process of volition and the result 'intellection' Individuals seeks Information for sheer curiosity' for new knowledge 'for solving problems' for decision making and for survival' in a meaningful way. In the light of library Automation it is the need of the hour to examine the system (i.e., Information retrieval process) frequently for better information organization and information retrieval. There is also a need to study the information seeking behaviours of the scientists, agricultural officials (i.e., Extension officers) and the target groups (i.e., farmers) and finally the outcome of their involvement through the eyes of public opinion (i.e., Press) in context of the development of agricultural output in the state. On the basis of experiences and responses to our queries, it is understood that the agricultural information available in the state is not sufficient and enough. Information relating to latest development in the field remains unknown and beyond reach to our scientists because of poor acquisition, storage, retrieval and dissemination mechanism of the library system in the state.

REFERENCES

1. Akoijam, Swarnalatha Devi and Lahiri, Ramansu. (1997). Information seeking behaviour of the Agricultural scientists in Manipur. IJILIS 10 (1-2), pp.60-67.
2. Phadnis, S.P. Shoab, A and Vijaya Lakshrmi, N.R. (1971). Training of agricultural scientists in information use. Journal of Library and Information science 2(2), pp. 151-181.
3. Raina, Rohan (1983), Libraries in agricultural research communication. Lucknow Librarian 15 (2), pp. 75-79.

USE OF COMPUTER IN AGRICULTURAL LIBRARIES IN HYDERABAD

INTRODUCTION

Computer applications in the field of Library and Information Science have been one of the major factors attracting the attention of the information professionals throughout the world. The emergence of microcomputers and the software available for the same have made computer application economically viable.

The library world is being acclimatized to computer environment in both library routines as well as information storage and retrieval. Computerization has retrieved the pressure of work load in libraries to a great extent. It offers efficient services and opens a new era in bibliographic control. It also provides access to required databases in the country and abroad. The computer is capable of storing information for longer period of time in minimum space in the electronic-readable form. This information is capable of being recalled with ease and utmost speed, as and when warranted.

AUTOMATION OF LIBRARIES

The facilities intended to be provided by automation are as follows:

1. Library house-keeping operations.
    - Acquisitions
    - Cataloguing
    - Serials Control
    - Circulation Control
2. Information Retrieval Approaches
    - In-house Catalogue
    - On-line Access
    - CD-ROM Discs

3. Library Network and Networking
4. Office Automation
   - Word Processing Applications
   - LAN (Local Area Network)

AGRICULTURAL LIBRARIES IN HYDERABAD

List of agricultural libraries in Hyderabad according to year of establishment and main source of finance are:

Agricultural Libraries in Hyderabad

Sl. No.	Name of the Library	Year	Finance
1	Andhra Pradesh Agricultural University	1964	State
2	Central Plant Protection Training Institute (CPPTI)	1969	Central
3	National Bureau of Plant Genetic Resource(NPBR).	1985	ICAR
4	National Institute of Agricultural Extension Management (MANAGE)	1991	Central
5	Directorate of Rice Research (DRR).	1970	ICAR
6	Central Research Institute for Dryland Agriculture (CRIDA).	1985	ICAR
7	Directorate of Oilseeds Research (DOR)	1967	ICAR
8	International Crops Research Institute for Semi-Arid Tropics (ICRISAT)	1972	Internet
9	National Research Centre for Sorghum (NRCS)	1987	ICAR
10	National Academy of Agricultural Research Management (NAARM).	1978	ICAR

As per the information given in the above table A.P. Agricultural University Library is the oldest and it has maximum number of collection.

OBJECTIVES OF THE STUDY

The study was conducted in order to find out

1. Number of Agricultural Libraries using computers.
2. Purpose of using computers in Agricultural Libraries in Hyderabad.
3. Willingness for using computers, if any library is not using.

LIMITATIONS OF THE STUDY

The study is limited to the survey of the agricultural libraries situated in Hyderabad.

ANALYSIS

The analysis of the data is presented and percentages are used for interpretation of the data.

Utilization of Computers

Sl. No	Response	No. of Libraries	%
1	In Use	7	70
2	Not In Use	3	30

The above table shows that 70% of libraries are using computers and 30% of libraries are not using computer.

Purpose of Using Computer (Presently Used)

Sl.No	Purpose	No. of Libraries	Reply	%
1	Storage and retrieval of Agricultural Information	7	7	100
2	Reference Services	7	2	29
3	Clerical and Administration	7	5	71

The above table reveals that 100 percentages of libraries are using computers for storage and retrieval of agricultural information. 71% of libraries are using computers for Clerical and Administration and only 29% of libraries are using computers for reference service.

Willingness to use Computers (Presently Non-Users)

Sl. No	Willingness	No. of Libraries	%
1.	Willingness	7	100
2.	Not willingness	-	-

Only seven libraries are not using computers and they are willing to use computers, if provided.

FINDINGS

The following are the findings of the present study.

1. Majority of agricultural libraries are using the computers.
2. Computers facilitated libraries are presently using computers only for the purpose of storage and retrieval of agricultural information.
3. 71 % of libraries are using computers for Clerical and Administrative work.
4. Majority of libraries are not using computers for Reference Services.

5. Presently computers non-facilitated libraries are also willing to use computers.

SUGGESTIONS

1. The library staff working in the Agricultural Libraries of Hyderabad is willing to use computers. So the management should take necessary steps to purchase computers for their libraries which are not purchased so far.
2. Sufficient and adequate staff (both professional as well as non-professional) to be provided to the Agricultural Libraries as per standards.
3. ICAR should provide necessary finance to automate of agricultural Libraries.
4. Training programmes should be conducted for staff to know the latest computer techniques.
5. ICAR should take steps to design Agricultural information networks for Hyderabad.
6. User orientation programmes should be conducted for readers to use computers information system.

REFERENCES

1. Gopinath, M.A. and Rama Reddy, E, ed. (1996). Information access through networks. Booklinks Corporation, Hyderabad.
2. Mittal, R.C. (1984). Library Administration : theory and practice. Metropolitan Book Co., New Delhi.

SURVEY ON THE INFORMATION NEEDS OF FARMERS IN NELLORE DISTRICT (A.P.)

INTRODUCTION

In the third five year plan much emphasis was given for communication of new technologies to farmers, but it failed due to holistic view reaching the farmers and consequent impact. India being a developing country with wide diversity in climate and farming systems including farmers, centralized systems of communication such as Radio, T. V., News paper could not play the expected role.

Significance and role of agricultural information for agricultural development was realized after independence when the agriculture changed fast and new techniques were being used in Indian agriculture. Farm information was regarded as one of the inputs in agricultural production and farm information was made available through various sources including print media.

Among mass media news papers, radio and T. V. are the most commonly used media. News papers are primarily a means of disseminating news and influencing the public on the current events. These days television is greatly used for educating, informing and entertaining the masses. The purpose of any communication medium is to put the ideas, make the meaning clear to get them accepted. The motivated people will adopt and practice them. Television can, therefore, be assumed to be a powerful medium.

Farmers are the ultimate users of the agricultural research findings through proper extension media. The farming category in the information communication process is likely to include illiterate as well as the literate strata. The extension workers and rural libraries have a significant role to play for communicating new ideas and practices to the farmers. For progressive educated farmers the best means of communicating research results for the library is through publications.

OBJECTIVES OF THE STUDY

The objectives of the study are as follows:

1. Finding out various types of information sources of farmers regarding seeds, fertilizers, pesticides and marketing.
2. Knowing the various ways through which the farmers seek their information.
3. Analyzing the role of Radio and T.V. informing the latest developments to the farmers.
4. Assessing of farmers preference regarding news papers and farm magazines published in Telugu.

METHODOLOGY

The data required for the study was collected from primary as well as secondary sources. The primary data was collected through questionnaires by interview method. The secondary data was collected from district hand book and agricultural abstracts available with Chief Planning officer, Nellore.

SELECTION OF SAMPLES

The study was conducted in Nellore district of Andhra Pradesh during the year 1995. 150 educated farmers were selected from Nellore district for the purpose of this study. As the Nellore District is having three Revenue Divisions, 50 samples were selected from each Revenue Division.

PROFILE OF NELLORE DISTRICT

Nellore district has three Revenue Divisions consisting 46 Mandals which were categorized with effect from 25 May, 1985. The three revenue divisions are Nellore, Kavali and Gudur covering 12, 18 and 16 Mandals respectively. Soils in the district are Black soils, Red soils and Clay loamy soils constituting 23%, 43% and 34% of the total area respectively. The main rivers Pennar, Kandaleru and Swarnamuki drain the district by traversing through West to East of the District and flow into Bay of

Bengal. Normal rainfall of the district is 1041 m.m. which is more than State average by 141 m.m.

The total geographical area of the district is 32, 52, 006 acres of which 23.17% are under cultivation. 7.15 lakh acres are irrigated which form 78.95 percent of total cropped ares. The principal crops raised in the district are Paddy, Jowar, Bajra, Chillies, Groundnut and Tobacco.

India's principal rocket launching station SHAR of ISRO is located in the Island of Sriharikota in Nellore district. The well developed physical and financial infrastructure facilities, availability of adequate raw material offer good scope for the development of agriculture, its allied activities and the promotion of wide range industries in the district.

ANALYSIS OF DATA

Table 1: Farmer's Source of Information about Seeds, Fertilizers, Pesticides and Marketing

Sl. No	Source	Total Respondents	Responses	% of response total respondents
1	Friends	150	93	62
2	Relatives	150	87	58
3	News Papers	150	69	46
4	Radio	150	72	48
5	T.V.	150	39	26
6	Govt. Agencies	150	79	50
7	Wall Posters	150	9	6
8	Representatives of Private Company	150	6	4

The Table 1 reveals farmer's source of information regarding seeds, fertilizers, pesticides and marketing. Majority of the respondents i.e. 62 percentage are getting required information from friends, 58 percentage of respondents are getting required information from relatives and 50% of respondents are seeking from government agencies.

The above analysis concludes that farmers were getting agricultural information mainly from friends, relatives and government agencies.

Table 2: Source of Information from news papers & farm magazines

Sl. No	News Papers / Farm magazines	Total Respondents	Respondents	% of respondents by source total respondents
News papers				
1.	Eenadu	150	69	46
2.	Andhra Jyothy	150	26	39
3.	Andhra Prabha	150	-	-
Farm Magazines				
1.	Annadata	150	27	18
2.	Mapally	150	-	-

It is evident. from the above table, that majority (46%) of farmers are getting required agricultural information by reading *Eenadu,* and 39 percentage of respondents are reading *Andhra Jvothi,* 18 percentage of respondents are using *Annadata,* farm magazine as a source.

The above information shows that the farmers are getting agricultural information only from two news papers *(Eenadu* and *Andhra Jvothi).*

Table 3: Farmers and Air Broadcasts

Sl. No.	Reply	Respondents	% of Respondents
1.	Yes	63	42
2.	No	87	58
Total		150	100

The Table 3 shows that only 42% of farmers are hearing AIR broadcasts and remaining respondents are not hearing.

Table 4: Types of Popular Programmes

Sl. No	Programmes	Total Respondents	Respondents	% of No. Respondents
1.	Polampanulu	63	09	14.29
2.	Padipantalu	63	05	7.94
3.	Palleseemalu	63	56	88.89

The Table 4 shows that most of the farmers are hearing palleseemalu programme. Hence it is clear that the palleseemalu programme is convenient programme for farmers, because it is being broadcast in the evening times.

Table 5: Farmer's Programmes on T.V.

Sl. No.	Reply	Respondents	% of Respondents
1.	Yes	31	20.67
2.	No	119	79.33
Total		150	100.00

It is evident from the Table 5 that more than 75% of the farmers are not watching T.V. programmes. Only 20.67% of respondents are watching agricultural programmes on T.V.

Table 6 : Reasons for not watching T.V. Programmes

Sl. No	Reasons	Respondents	% of No. Respondents
1.	Non-availability of T.V.	102	85.71
2.	Non-availability of time	17	14.29
3.	Other reasons	-	-
Total		119	100

The Table 6 shows that percentage of the farmers not watching the T.V. programmes because of the non-availability. Very few respondents replied that they have no time to watch T.V. progrnmmes.

FINDINGS

1. Majority of the farmers are getting agricultural information regarding seeds, fertilizers, pesticides and marketmg mainly from friends, relatives and Government agencies.
2. It is clear from the analysis, that the respondents are reading only two news papers *(Eenadu* and *Andhra Jyothi)* for agricultural information,
3. Half of the respondents are not able to hear AIR broadcasts because of financial problems.
4. The analysis shows that the *"palleseemalu"* programme is convenient programme for farmers, because of its being broadcast in the evening time.
5. Nearly 80% of the farmers are not watching T.V. programmes. The main reason for non-observation is because of the finance to buy T,V. sets.

SUGGESTIONS

1. Most of the farmers are not able to hear AIR broad- casts and watch T.V. programmes. This situation can be overcome by providing Radio and T. V, to villages.

2. All India Radio and Doordarshan should broadcast agriculture related programmes only in the evening times and not in the day time, as the farmers are busy with their work.
3. The Agricultural University should establish agricultural information communication centres for fulfilling the informtation requirements of the farmers.
4. There is an immediate necessity to build up the Rural Information Centres as integral part of the public library system with the help of Agricultural University and ICAR.
5. The Information and Public Relations Department and Agricultural Department should open information centres in every Mandal and screen regularly film on the latest developments in agriculture.
6. The Government should start Krishi Vigyan Kendras and Farmers Training Centres in every district.

REFERENCES

1. Bhagat. Rekha and Ramachandran, C (1993). Trends in Agricultural communication research. *Communicator* XXVIII (4).
2. Gangadhara Rao, P (1986). Rural Information Centres. National seminar on assessment of public library development in India.
3. Mohan, Surendra (1997). Agricultural Information Systems and Services. *Journal of Library and Information Science* 2(2).
4. Nanjappa, D and Ganapathy. News paper reading habit of farmers of Bangalore district in Karnataka. *Indian Journal of extension education* XXII (1& 2).
5. Raina, Roshan (19983). Libraries in agricultural research communication. *Lucknow Librarian* 15(2) 1983.

PREPARING A RESEARCH ARTICLE: SOME USEFUL GUIDELINES

Planning and Observation

"Failing to prepare is preparing to fail."

Thorough preparation, as such, is a must before writing a scientific paper.

Type of presentation

The first step towards preparation is to decide the form of your paper that is whether it is going to be a review article, an original article, or a letter to the editor.

Selecting the Journal

Select a journal which has a wide audience and which is closely related to your field of study. Ascertain its accessibility, check the members of the editorial board and determine if they are the stalwarts in their fields. Determine the journal's impact factor and try to find out its acceptance rate. The impact factor is calculated by dividing the number of current citations to articles published in the two previous years by the total number of articles published in that period. Have a look at the style and format of the journal; find out the time it takes to publish; and, finally, if it bills the author for pages. For the beginners, it is advisable to go for monthly peer-reviewed journals and when you have got a few publications, you can go for the international journals having high impact factor.

Planning of Author's Name

Decide on the authorship in the beginning itself if the article is going to be a joint contribution. It would be an intellectual dishonesty if you include the names of persons who have not made any contribution as authors. The framework of "honorary authorship" should be avoided.

Certain pre-requisites for designing a paper

Before you start writing a paper, have the following things in your mind.

1. Write as per the format described in the journal, that is your writing style, titles and formatting style should go according to the journal you are writing for. Also, find out if online submission of the manuscript is allowed or not.
2. Get the appropriate ethical, statistical and copyright clearance for the paper beforehand. It is always better to collect and prepare illustrations, tables, graphs and photographs before you start writing the paper.

Reasons for writing research articles

- Career development and promotions.
- To enhance the image and prestige of the University/Institute.
- To share experiences and findings.
- Professional development.
- Hobby.
- Competition among teachers and scientists.

Sources to gather material to write research articles

- Referring Journals.
- Conducting research work.
- Supervising the research students.
- Discussing with teachers, seniors & colleagues.

- Attending Seminars, Workshops etc.
- Conducting projects.

Writing a Research Article

Writing a paper requires utmost tact and care.

We should keep in mind certain DO's and DON'Ts while writing a research paper:

1. Do not hesitate to rewrite any number of times since there is no short cut to produce something which consumes the time of others to read.
2. Do not fudge data or mislead, since truth will ultimately come out.
3. Do not duplicate and camouflage the scientific substance through literary flourishes.
4. Do start writing the paper before the experiments are completed since writing is a continuous process and you embellish the quality of your writing with the passage of time.
5. Do not include tables and figures more than necessary.
6. Do not write in first person. Instead use the words " The Researcher" or "The Investigator".

DECIDE YOUR FORMAT

There are three basic forms of research articles. Choose the category to which your article belongs before you start writing:

Full-length research articles

These articles cover comprehensive research carried out over a length of period. They follow "IMRAD" format -Introduction, Methods, Results and Discussion.

Short communications

The information is presented as a single section with tables and figures.

Rapid communications

Present "hot" findings in a brief communicational format.

In a nutshell your paper should contain the following.

Experimental Process	**Section of paper**
What did I do (brief)	Abstract
What is the problem	Introduction
How did I solve the problem	Methods and Materials
What did I find out	Results
What does it mean	Discussion
Who helped me out	Acknowledgements (Optional)
Whose work did I refer to	References
Extra Information	Tables and Figures

LET US EXAMINE THESE STEPS IN DETAIL

Title

The readability of your article to a large extent depends on the title. Select a short, simple and eye-catching title which goes well with its contents.

Abstract

Your chances of getting your paper published depend on the Abstract. In a capsule, it should cover the entire information like the aim of the study, the methodology used, results obtained and conclusions drawn. An abstract between 150-250 words is considered to be "good". Devote a considerable amount of time for preparing an abstract.

Introduction

Should not be too long as seen in most of the cases. Explain the significance of the study by spelling out hypothesis. Also, explain the reason for undertaking the research.

Methods and Materials

Explain the following clearly in this part:

- Type of study/approach you have undertaken – descriptive, ethnographic, narrative/holistic with details like location, date and time of carrying out experiments.
- Sampling design and methods used for data collection.
- Statistical procedures used for analyzing data.

Results

Avoid using terms like majority, most of the, in some cases, significantly greater than, more or less, could be due to, etc. Explain the results with confidence and have faith in yourself. Do not hesitate to give negative results or unexpected findings if any. Ignore percentages or statistical figures when the number is small.

Discussion

It is usually a good idea to begin this section by giving the answer to the question you set out with. State the main findings clearly through flow diagrams and tables and, if possible, compare this data with other published information. Examine the ramification in detail. Do not, however, introduce new results in this part. Reserve this for your "major findings" unit.

Acknowledgements

There are no hard and fast rules for acknowledging somebody's help, be it that of a person, an institution or an organization. The investigator, as such, is at liberty to write this part in his/her own style. This unit is also not compulsory but a word of thanks will really go a long way in paying intellectual dividends to all those who helped you in carrying out your work.

References

The purpose of this section is to provide the full citation for articles referred to in the text. A complete reference includes all the authors' names, the titles of the articles, the journals' name, the volume number, page numbers and the year of publication. A wide range of styles is used for citing references in the text and bibliography. Check the journals' "instructions to authors" for information about the content and formatting references and write accordingly.

At the end of the paper, a list of references is provided in alphabetical order. In some cases citations appear in the text as numbers, usually a superscript, which then refer to a particular item in the reference list. It is compulsory for the author to check all the references thoroughly before citing them since simply copying them is likely to be error-prone. There are no rules to be followed in letter and spirit with regard to the number of references. Some journals nevertheless impose a limit to conserve space. A rule of thumb is not more than 6 references for a particular point and not more than 100 references per paper.

Tables and Figures

Do not include Tables and Figures more than necessary since they may divert the attention of your reader. Remember that an ignorant man collects the data whereas a wise man selects the data.

GUIDELINES OF MANUSCRIPT DESIGN AND BASIC RULES OF WRITING A RESEARCH ARTICLE

Sections	Main functions	Preferred style	Basic rules
Title	Indicates the entire matter, catches the reader's attention	Short, simple, clear and unambiguous	Avoid high flown language, make it catchy and avoid redundancy. Spend some time on finalizing the title.
Abstract	Explains the entire story in a capsulated form	Past tense and passive voice – no citations, tables, graphs etc.	Explain what was done and what was the outcome without introducing the theme.
Introduction	Introduces the topic, explains the terminology, the focus of the paper, the research objectives, hypothesis and limitations of the study	Use simple language for referring to established knowledge	Do not prolong this section by inserting redundancies Explain the significance of the study step by step in a logical manner.

Contd....

Sections	Main functions	Preferred style	Basic rules
Review of related literature	Cites the related research undertaken in the area concerned.	Past tense	Use state-of-the art references. Avoid irrelevancy
Methodology	Explains the 5 W's, namely Who, What, When, Where and Why of your study	Past tense	Explain everything in detail. Use authoritative tone and simple language.
Results	Gives summary of the results obtained through statistical tests in written as well as graphic forms	Past tense	Emphasize the most significant findings. Distinguish your work from that of the other
Conclusions and Recommendations	Reveals the important research findings and future implications	Past tense for conclusions-Researcher's style for Recommendations	Do not recapitulate results but make strong statements. Do not hide unexpected / negative results.

Editing and Revising

Editing and revising involves the following steps in the order:

1. *Make major alterations:* Fill in the gaps, change the order of contents to present the material in an acceptable form.
2. *Fine-tune the style:* Refine the text and make it free from grammatical and spelling errors. Get it proof read by others.
3. *Get feedback:* Get a feedback of your document from your peers. Make modifications as per necessity and requirement.
4. *Format the document:* Make your document attractive and easy to read.

Self-revising involves revision and modification of matter starting from a global perspective, coming down to paragraph content and finally down to sentence-level editing.

Global

Check whether there is cohesive, logical and sequential presentation of ideas from one topic to another.

Paragraph

Check the order of paragraphs to find out if there is a smoother transition between them.

Line editing

Eliminate superfluous and colloquial phrases; avoid redundancy; read the paper aloud; and check and recheck grammatical and spelling errors.

Checklist before the Final Submission of Your Document

1. Check all your sources are cited correctly.
2. Check the number of tables and figures.
3. Ensure that the authors' names are spelt out correctly.

4. Check the line spacing.
5. Check you have given your document for proofreading at least twice to two different persons.
6. Re-read the document slowly for the last time before giving it for final publication.

Submitting Manuscript and Follow-ups

As soon as your manuscript is ready, start the process of sending it for publication. Remember that you will have to go through the following procedure before getting your document published.

Identify the mode of submission

Prefer online submission if this option is available as it has its own advantages over the normal postal submission.

Prepare a covering letter

The covering letter should be simple, short and effective addressing the editor by name as it has its own impact. Do not forget to write the title of the article, authors' name and date.

Submit the manuscript to the Editor

Follow the instructions to authors scrupulously. Some journals ask for a 'pre- review", i.e., a letter indicating the content of the article to determine its worth and credibility. List out the possible reviewers in this case.

Handling the reviewer's comments

Acceptance of manuscripts on first submission is rare. Do not get discouraged if this happens to you. Modify your manuscript with tact if the editor so desires. Answer every concern of the reviewer and indicate where the changes were made. Approach the revision using global, paragraph and line editing strategy as discussed earlier. While resubmitting, mention in your cover letter that this a revised version. Do not submit your manuscript to more than one journal at a time.

Check the proofs

Your manuscript, if accepted for publication, will be sent to you for proofreading. The proofs may be sent via email or hard copy. Check the proofs with utmost care and patience.

Celebrate

Thomas Alva Edison says that success is 99% perspiration and 1 % inspiration. After putting immeasurable amount of hard work and perspiration, you have now ultimately reached your goal, i.e., publication. Enjoy and rejoice the occasion.

AVOID REJECTION

Be thorough with the following in order to avoid rejection.

Be well conversant with the editorial process

Your article will be reviewed by the editor, associate editor and reviewer. They will evaluate your documentation in the following areas:

- Significance and relevance of the topic in the present-day scenario.
- Literature review.
- Sequential and logical presentation of ideas.
- Conclusions/Recommendations that have emerged out of your study.
- Clarity and consistency of writing.

Do not feel disheartened if the editor asks you for any of these modifications. Resubmit your manuscript as per the requirement of the publisher.

AVOID PITFALLS

The most important reasons for rejecting a research article are:

Area	Reasons for rejection
Topic	Irrelevant and restricted.
Lack of Innovation	Repetition of the work done earlier.
Focus	No cohesion between the topic, objectives and conclusions.
Methodology	Weak, not clearly spelt out. Data with insignificant sample number and flaws.
Style	Detached and unfocussed.

Given below are a few tips for improving the probability of acceptance of your research article.

What to do	How to do
Take a reader's view	Imagine yourself to be the reader while you write.
Simplify	Be yourself. Use short sentences and simple examples to explain complex methodology.
Avoid	Redundancy, repetition and over-explanation of familiar techniques and terminology.
Make strong statements	Use "We/I concluded" instead of "It may be concluded"
Be self-critical	Be open to unexpected conclusions and acknowledge the work of others.

To assist the writer in preparing a quality manuscript, most journals publish a guide to article preparation or recommend a particular guide to follow.

To summarize, here are some suggestions for improving a manuscript:

- Use simple language whenever possible and carefully explain difficult concepts or terminology;
- Short sentences are preferable as they are clear, concise, jargon-free and do not distort the language;
- Avoid unnecessary repetition and use the active voice wherever possible;
- The old saying "Say what you are going to do. Do it. Then say what you have done," provides a useful model for overall organization of the manuscript and its subsections. The first anticipates, the second elaborates while third summarizes and takes stock;
- Brevity often goes with clarity. Most manuscripts should be limited to a single major issue or argument.

HOW TO PREPARE A PROJECT PROPOSAL

This is a modest attempt considered to be of prime importance to help the young scientists in pursuing their research thoughts by providing the required information to prepare a research project proposal and sent it to a funding agency. Before writing a project proposal, a person should self assess his/her technical capability in the field he/she is going to propose. Writing good projects is an intuition to a few scientists while many of them sharpen their abilities and acquire the talent through contemplation, perception and interaction with specialized groups. Talent with knowledge, imagination, clarity in conception and expression etc. make a person write a good project.

Individuals or groups working in an institution/organization are eligible to apply for funds. Some funding agencies extend the opportunities to the retired scientists also. They even include the private R&D labs and also NGO organizations. It is undoubtedly the areas of research that get support and not the individuals or organizations. Generally the funding agencies accept proposals from the Heads of Institutions but not directly from the Principal Investigators. You should never forget that your project proposals are reviewed by experts and based on their recommendations the project will be sanctioned.

Once an idea is conceived, list out the funding agencies and know about their thrust areas of research. The project proposal should suit well to any discipline of the funding agency. Funds are available for innovative research rather than repetitive/confirmative research programmes. The thrust areas of the funding agencies will be periodically reviewed and changed with emerging needs. So keep abreast with the changing scenario.

Identification of a good research problem means completing half of the research work. Normally ideas prop up during feedback given by farmers, extension personnel discussions held at research stations, zonal research extension and advisory council meetings, state level technical programmes and at national workshops. A through literature search helps you identify the research gaps where you can make a good contribution. Problems of your locality with your material would be unique and worth supporting for funding. You should not forget to quote your labs work and involvement in the proposed field. Avoid unnecessary references, which otherwise speak of low level of your confidence.

After identifying the research gaps finalize the title and list out the objectives by planning a few full pledged experiments, which can answer many questions within a short period. Too many objectives may be disgusting. Do not touch upon the controversial topics while preparing proposals. Your intention should not be the procurement of equipment. Hence limit the number of equipment only to the necessity. Discuss with experienced staff and include one or two reputed scientists' names that who already got projects. Split the plan of work amongst the collaborators (inter disciplinary or inter-institutional) based on the capacity to complete within stipulated time. It is always better to prepare proposals after some preliminary studies. Plan for small projects with an outlay of less than Rs. 10 lakhs initially.

The methodology section should be prepared in detailed manner. For each objective the background and the conceptual basis should be explained outlining the hypothesis be tested or observations to be made. Where ever possible adopt statistical analyses using figures and graphs practical relevance of the research in terms of development of technology. Direct and indirect benefits especially quantification need to be highlighted. Remember that failures are the stepping-stones to success. If your proposal is rejected, write again till you succeed.